# *Are We Alone In The Universe?*

**The Search for Extraterrestrial Life**

by
Miles Kepler

Miles Kepler

Copyright 2024 Miles Kepler. All Rights reserved. No part of this publication may be reproduced without consent of the author.

"I'm sure the universe is full of intelligent life. It's just been too intelligent to come here."

**— Arthur C. Clarke**

Miles Kepler

**Table of Contents**

Chapter 1: The Allure of Extraterrestrial Life

Chapter 2: Conditions for Life: The Goldilocks Zone

Chapter 3: Unraveling Mars: Close to Home

Chapter 4: Moons of Mystery: Europa and Enceladus

Chapter 5: The Exoplanet Revolution

Chapter 6: The Search for Signals: The SETI Initiative

Chapter 7: Life's Extremes: Lessons from Earth

Chapter 8: Sci-Fi and the Search for Extraterrestrial Life

Chapter 9: The Consequences of Discovery

Chapter 10: The Future Awaits: Continuing the Search

*Introduction*

Have you ever gazed up at the night sky, marveling at the countless stars twinkling in the darkness, and wondered if we're truly alone in the universe? This question has captivated human imagination for millennia, driving us to explore the cosmos and search for signs of life beyond our blue planet.

"Are we alone?" This simple yet profound query has shaped mythologies, inspired scientific breakthroughs, and fueled countless debates. From ancient civilizations that saw gods in the celestial bodies to modern-day scientists peering through powerful telescopes, our species has been on a relentless quest to uncover the secrets of the universe.

As you embark on this journey through "Life Beyond Earth: The Search for Extraterrestrial Life," prepare to explore the fascinating intersection of science, philosophy, and human curiosity. We'll traverse the timeline of humanity's cosmic fascination, from the first stargazers who mapped the night sky to the cutting-edge technologies probing distant planets for signs of life.

This book is not just about the search for alien life; it's a reflection on our place in the vast cosmic tapestry. It's about the questions that drive us forward and the implications of potentially finding life elsewhere in the universe. What would it mean for our understanding of existence? How might it reshape our cultures, beliefs, and sense of self?

As we venture into the unknown, we invite you to join us in contemplating these profound questions. Let your imagination soar as we explore the possibilities that lie beyond our terrestrial boundaries. Are we on the brink of the greatest discovery in human history? Or are we truly alone in this vast cosmic ocean?

Buckle up for an awe-inspiring journey through space and time. The universe awaits, full of mystery and wonder. Are you ready to explore?

Miles Kepler

# Chapter 1: The Allure of Extraterrestrial Life

The night sky, filled with sparkling stars and swirling galaxies, has always sparked our imagination. For thousands of years, our ancestors looked up in awe, brimming with curiosity about what was out there, beyond their earthly lives. Ancient cultures created stories that intertwined the divine, the celestial, and the idea of extraterrestrial life, building a deep bond with the universe that still resonates with us today.

Take the Egyptians, for example. Their pyramids stand as magnificent reminders of how they viewed the afterlife and the cosmos. These grand structures were aligned with the stars, especially the constellation Orion, which was linked to Osiris, the god of the afterlife. They believed that pharaohs would rise to the heavens and join the gods among the stars. This belief created a powerful connection between life on Earth and the mysteries of the universe. For them, navigating by the stars was not just a practical matter; it was a deeply spiritual act, showing how ancient people looked to the skies for guidance, meaning, and understanding of their own existence.

Meanwhile, the Mayans meticulously tracked celestial events with their complex calendars, recognizing the importance of cycles in both time and life. They looked to the heavens, interpreting the movements of planets and stars as messages from the gods. These celestial alignments dictated when to plant crops and when to hold religious ceremonies, weaving the stars into the very essence of their civilization. Their belief in the cyclical nature of life echoed the idea that our existence on Earth was part of a grand cosmic order, which might include other forms of life among the stars.

Sumerian mythology gives us yet another intriguing view. The Anunnaki, a group of powerful deities, were believed to have come down from the heavens to share knowledge and help build civilization. These stories show how early societies tried to make sense of their lives in the context of the vast universe. The Anunnaki were often depicted as mighty beings shaping human destiny—an interplay of fear, respect, and hope that still captivates our imaginations today.

Astrology played a big part in shaping beliefs about life beyond our planet. Ancient civilizations closely followed the movements of celestial bodies to predict everything from weather changes to personal fortunes. The Zodiac, with its twelve signs, offered a way for people to interpret their lives and their connection to the cosmos. The idea that stars could influence earthly events reinforced the belief that we are not just isolated beings; we are all part of a larger cosmic story.

As humanity entered the Age of Enlightenment, the scientific revolution began to unravel many of these ancient beliefs. With the invention of telescopes and the rise of astronomy, stargazing shifted from mystical wonder to scientific observation. The universe, once filled with gods and myths, started to reveal itself as a vast expanse of stars and planets, possibly housing life. Still, the intrigue of extraterrestrial life persisted; it simply transitioned from mythology to scientific exploration.

Fast forward to today, when our interest in discovering extraterrestrial life has been reignited by leaps in technology and our growing understanding of the universe. The launch of space telescopes like Kepler and TESS has uncovered a treasure chest of exoplanets—worlds orbiting distant stars that might have conditions suitable for life. In an age where we can capture images of far-off galaxies and detect the faint glow of atmospheres on distant planets, the pressing question of whether we are alone in the universe feels more urgent than ever.

The fascination with life beyond Earth is no longer just a myth or a dream. It has turned into a real pursuit, with scientists and curious minds alike peering into the cosmos, driven by an unquenchable thirst to discover what—or who—might be out there. Each discovery has the power to change how we view life, encouraging us to reflect on our own existence and the endless possibilities that the universe may offer.

Consider the recent efforts of astrobiologists who study extreme environments here on Earth—places like deep-sea vents and acidic lakes—to understand how life might survive in harsh conditions elsewhere in the universe. The discovery of extremophiles, organisms that thrive in environments once thought uninhabitable, has broadened our understanding of what it takes for life to exist. This leads to exciting possibilities: could life be lurking in the hidden oceans of icy moons like Europa or floating in the thick atmosphere of Venus?

The scientific community has also taken on the search for extraterrestrial intelligence (SETI), actively listening for signals from advanced civilizations. The thought that we might not be alone—that there could be other intelligent beings out there—fills us with both excitement and wonder. What would it mean for us if we received a message from another world? It would challenge our understanding of life, intelligence, and our role in the universe.

Public interest in extraterrestrial life continues to grow, fueled by how popular culture portrays aliens—from the friendly E.T. to the scary creatures of science fiction. Movies, books, and TV shows give us a way to explore our fears and hopes about the unknown. They invite us to think about the possibilities of contact, coexistence, or even conflicts with beings from other worlds. These stories significantly shape our culture and ignite our imagination, inspiring us to ponder the big questions of the universe.

A shared desire to connect with whatever exists beyond our planet is clear in the increasing number of projects aimed at sending messages into space. Initiatives like the Voyager Golden Record, launched into the cosmos in 1977, symbolize humanity's message in a bottle, showcasing sounds and images that represent who we are and what our world is like. This effort to reach out to the universe reflects a longing for connection, a wish to be heard amid the vastness of space.

As we navigate the complexities of our lives, the allure of extraterrestrial life remains as strong as ever. It invites us to think about where we came from, why we are here, and the limitless possibilities that lie beyond our planet. The quest to know if we are alone in the universe might be as much about understanding ourselves as it is about exploring the stars. It calls on us to face our own humanity and expand our horizons, encouraging us to gaze up at the night sky with not just curiosity, but with a determination to find answers.

Ultimately, our fascination with life beyond Earth reflects our deep-seated desire to understand the cosmos and our role in it. As we continue to explore and gain knowledge, we are reminded that looking outward may lead us to valuable insights about ourselves. The stars are not just distant points of light; they hold the questions that have lingered in human hearts for centuries, urging us to seek, to wonder, and, in the end, to connect with the universe in ways we are just starting to grasp.

## The Impact of Literature and Film

Literature and film have long provided us with a powerful way to explore the intriguing idea of extraterrestrial life. Whether it's the excitement of a close encounter or the anxiety of an alien invasion, these stories have shaped our imaginations and reflected our shared feelings as humans. They don't just entertain us; they offer deep insights into our hopes, fears, and what it really means to be human in a universe that might be full of life.

Take, for example, H.G. Wells' classic story, "The War of the Worlds," published in 1898. This groundbreaking work is a cornerstone of science fiction, skillfully intertwining themes of imperialism with the dread that comes from the threat of an alien attack. The Martians are shown as merciless and advanced beings bent on conquering Earth, using terrifying machines and deadly weapons. This gripping tale not only fascinated readers but also echoed the worries of its time. The British Empire was at its peak, and the fear of foreign conquest struck a chord with many. By depicting the Martians as invaders, Wells captured the cultural concerns of his era, turning the alien into a reflection of our own ambitions and the vulnerability of human civilization.

Yet, Wells' story is more than just a frightening narrative; it reveals a complex relationship between fear and curiosity. Though they are terrifying, the Martians also spark a sense of wonder. They embody the unknown and the potential for discoveries beyond our comprehension. This duality shows how literature often explores the depths of our minds, bringing our biggest fears to light while also stirring our sense of wonder. Facing an alien presence pushes us to reflect on our own existence and to think about what it means to be human in a universe that might be full of life.

Now let's look to the movies, where stories about aliens continued to flourish. Films like "Close Encounters of the Third Kind," directed by Steven Spielberg and released in 1977, changed how we perceive the idea of alien life. Unlike the Martians of Wells' tale, Spielberg's aliens aren't malicious conquerors; they are mysterious visitors, inviting curiosity and awe. The film's memorable scenes—where regular people are drawn to a strange signal, leading to a stunning meeting with extraterrestrials—symbolize our deep desire for connection.

In "Close Encounters of the Third Kind," the encounter isn't framed as something to fear but as an opportunity for wonder. The film encourages viewers to think about peaceful interactions, filling us with hope that we might not be alone and that the universe holds more than just dangers. The climax, filled with beautiful music and breathtaking visuals, suggests that meeting alien life could elevate our understanding of our place in the universe. This isn't just a tale of man against monster; it's about the potential for communication and understanding, reminding us that we are part of something far greater than ourselves.

The mix of fear and wonder is a theme that keeps coming back in movies about extraterrestrial life. Films like "Alien" and "The Thing" dive into the darker, primal fears tied to the unknown. In these stories, aliens not only pose a physical threat but also symbolize deeper anxieties about invasion, contamination, and losing our identity. The sheer horror of facing an alien force makes us question the very essence of humanity and what it means to share existence with other beings, whether they come from our own world or the stars.

These portrayals have real psychological implications that go beyond just entertainment. They offer a way for society to process its fears and dreams. As humans, we are naturally drawn to stories that challenge our views of existence. Literature and film give us a chance to explore the possibilities of life beyond Earth, reflecting our own moral dilemmas and philosophical questions. Engaging with these narratives leads us to wrestle with ideas about consciousness, identity, and the ethical choices we make, both on our planet and beyond.

It's also important to recognize how literature and film have sparked public interest in the quest for extraterrestrial life. As tales of alien encounters become part of popular culture, they inspire real scientific exploration. The fascination these stories evoke has motivated many to pursue careers in astronomy, astrobiology, and aerospace science. The Apollo missions, the Mars Rover, and the ongoing search for exoplanets are all, in part, responses to the dreams and fears brought to life in books and movies.

Consider the "Star Wars" franchise, which not only reshaped science fiction but also created a rich universe filled with diverse alien species and complex interstellar politics. The stories in "Star Wars" prompt viewers to think about cooperation, conflict, and the ethical implications of power. As we follow characters like Luke Skywalker and Princess Leia, we delve into themes of heroism and sacrifice against overwhelming challenges. This universe, brimming with culture and complexity, ignites our imagination and encourages us to contemplate what it might be like to live alongside a multitude of intelligent beings, each with their own unique histories and values.

Exploring the realms of science fiction blurs the line between reality and imagination. The stories we engage with shape how we view extraterrestrial life, bringing forth both hope and fear. They capture our dreams of exploration, our thrill at discovery, and our anxiety about the unknown. The emotional tug-of-war between these contrasting feelings is a key part of the human experience—reminding us that when we gaze up at the stars, we're not just looking for life beyond our planet; we're also searching for meaning in our own lives.

Furthermore, the influence of literature and film reaches into the scientific community, where researchers often find inspiration in creative narratives. Ideas that emerge from fiction resonate within the walls of laboratories and observatories. Concepts like the Fermi Paradox, which questions why we haven't encountered extraterrestrial civilizations despite the vastness of the galaxy, reflect themes explored in countless stories. Just thinking about this paradox shows how fiction can spark real-world exploration, pushing scientists to devise ways to search for life in the universe.

The success of documentaries like "Cosmos: A Spacetime Odyssey" highlights this connection. By marrying scientific knowledge with captivating storytelling, these programs not only educate us but also inspire us to think critically about the universe. They bridge the gap between scientific exploration and the imaginative tales found in literature and film, inviting viewers to wonder about our role in the cosmos while staying grounded in reality.

As we think about the impact of literature and film on our understanding of extraterrestrial life, it's clear that these narratives are more than simple entertainment; they're cultural creations that capture our deepest wishes and fears. They challenge us to face the unknown and reconsider our beliefs about life beyond Earth. Whether through the fright of alien invasions or the excitement of meeting benevolent beings, these stories resonate deeply within us, shaping how we perceive the world and fueling our desire to learn more.

The fascination with extraterrestrial life remains a strong force in our society, showcasing the creativity and imagination that define us as humans. As we keep exploring the universe and all it has to offer, we should stay open to the lessons offered by literature and film. They're not just stories; they're calls to engage with the cosmos, to ponder our place in the universe, and to embrace the wonder waiting just beyond our planet. From the chilling echoes of Wells' Martians to the uplifting harmonies of Spielberg's visitors, these narratives remind us that the search for extraterrestrial life is just as much about understanding ourselves as it is about uncovering the mysteries of the universe. It's a journey that invites us to look beyond the stars and, in doing so, to discover our own humanity.

## Scientific Inquiry: The Drake Equation

When we look up at the night sky, filled with countless stars and planets, it's hard not to wonder: Are we alone in this vast universe? This question has stirred the imaginations of scientists, philosophers, and dreamers alike, sparking a sense of curiosity that has lasted for generations. With the universe stretching beyond what we can fully grasp, the search for extraterrestrial life has turned into an exciting and sometimes overwhelming adventure. One of the most intriguing tools we have in this quest is the Drake Equation. It shines a light on the possibilities of alien civilizations in our galaxy and gives us a structured way to explore the question of life beyond Earth.

The Drake Equation, created in 1961 by astrophysicist Frank Drake, is far more than just a set of numbers. It opens the door to a deeper understanding of our place in the cosmos. This equation helps us examine what conditions are needed for life to exist elsewhere and guides scientists as they search for answers to whether we are indeed alone. The equation breaks down into several essential pieces, each representing a factor that contributes to the likelihood of intelligent life out there. These pieces include the rate of star formation, the fraction of stars that have planets, the number of those planets that could potentially support life, and the chances of life developing on them, among others.

To grasp the importance of the Drake Equation fully, we need to take a closer look at these components, as each one plays a vital role in shaping our understanding of the universe's potential for life. The first factor—the rate of star formation—sets the stage for everything else. Our Milky Way galaxy is like a busy city of stars, with new stars being born at a rate of about one to three each year. This stunning number shows just how many stellar bodies could possibly host life-supporting systems. If we consider that every star might have its own planetary system, the possibilities for discovering life become even more exciting.

Next, we look at the fraction of stars that have planets. This idea was once debated, but thanks to advances in technology, we've learned a remarkable truth: planets are everywhere in our galaxy. Missions like NASA's Kepler Space Telescope have identified thousands of exoplanets, showcasing a dazzling variety in size, makeup, and distance from their stars. With estimates suggesting that nearly every star has at least one planet, the conversation about the potential for life really heats up.

Now, let's consider the number of planets that could support life. This brings us to the "Goldilocks Zone," where conditions are just right for life as we know it—neither too hot nor too cold. These are the planets where liquid water can exist and all the necessary ingredients for life are present. Our Earth sits snugly within this zone. As we discover more exoplanets that fit this description, the chance of finding life becomes even more tempting.

But we can't stop there. To really understand the potential for intelligent life, we also need to think about how likely it is for life to develop on those habitable planets. This part can be tricky since a variety of conditions must be just right for life to take root. The early Earth went through many chaotic and transformative events, from volcanic eruptions to asteroid impacts, leading to the complex web of life we see today. The possibility of life emerging elsewhere remains a mystery, but the discovery of extremophiles—organisms that thrive in harsh environments on Earth—shows us that life can be surprisingly tough and adaptable.

The equation then moves to the next factor: the fraction of planets with life that may go on to develop intelligent beings. This introduces the realm of evolution and the many paths it might take. Evolution is a complex journey influenced by countless environmental factors and chance events. The emergence of intelligent life on Earth is a unique story, prompting us to wonder if similar stories are unfolding on other worlds. Could a distant planet be home to advanced beings, or are we merely an outlier in the grand narrative of evolution?

Finally, the last parts of the Drake Equation focus on how long civilizations last and whether they can communicate across the vastness of space. How long can intelligent civilizations endure, and how many can send signals into the cosmos? This raises questions about societal progress, technological growth, and the potential for self-destruction. Human history has seen both incredible triumphs and serious threats, making us ponder the fate of other civilizations that might come and go in far-off star systems.

As we dive into the Drake Equation, it becomes clear that each piece is connected, weaving together a larger picture that influences our understanding of the universe and our role in it. Despite its mathematical beauty, the equation is not without its debates. Many of its variables carry a lot of uncertainty, sparking discussions among scientists and enthusiasts alike. Some argue that these variables are too speculative, while others believe they offer a valuable roadmap for exploration.

This ongoing conversation isn't just a matter of science; it has deeper philosophical implications. The Drake Equation pushes us to confront our thoughts about life, intelligence, and the universe as a whole. It encourages us to consider the idea that life, in its many forms, might be spread throughout the cosmos, waiting for us to find it. This challenges the long-held belief that humans are the center of it all, nudging us to rethink what it means to be alive and aware in such an expansive universe.

The implications of the Drake Equation reach far beyond science; they touch on deep questions about existence and the human experience itself. If we accept that the universe is rich with potential for life, we must also face the ethical and philosophical responsibilities that come with this realization. The possibility of discovering extraterrestrial civilizations raises profound questions about our role as caretakers of our planet and the broader impact of our actions in the universe.

Additionally, the quest to answer the questions posed by the Drake Equation inspires scientific breakthroughs and exploration. Projects like the SETI (Search for Extraterrestrial Intelligence) initiative highlight our determination to find signals from distant civilizations, fueled by the hope that we aren't alone. The advancements in technology, space missions, and telescopes designed to explore our galaxy reflect a collective commitment to answer the age-old question of whether we have cosmic neighbors.

    The Drake Equation has also made its way into popular culture, sparking a wealth of stories that explore themes of alien life, communication, and the possibilities of connection across the universe. Science fiction, in particular, has become a powerful way to examine what the equation means, offering imaginative tales of what might happen if we were to meet beings from other worlds. From friendly visitors eager to share knowledge to hostile invaders, the narratives we create reveal our hopes and fears, shaping how we understand the universe's potential.

    Beyond its scientific roots, the Drake Equation embodies the spirit of curiosity and wonder that drives us as humans. It captures our desire to explore the unknown and seek answers to the questions that have lingered throughout our history. The equation serves as a reminder that, in the grand scheme of things, we are just one thread woven into the fabric of the cosmos. As we search for extraterrestrial life, let's keep our hearts open to the mysteries waiting for us and embrace the journey of discovery ahead.

As we think about the components of the Drake Equation, we become part of a wider story that crosses the boundaries of science and taps into philosophy and existential thought. The equation bridges the gap between what we can prove and what we can imagine, challenging us to dream of possibilities while staying grounded in the reality of our universe.

The pursuit of extraterrestrial life, framed by the Drake Equation, goes beyond an academic exercise; it reflects a deep exploration of what it means to be human in a universe that can be both intimidating and awe-inspiring. As we gaze at the stars above, let's remember that we are on a timeless quest, one that could change how we see life, existence, and the endless possibilities that lie beyond our own world. The Drake Equation encourages us to move forward with bravery, curiosity, and a shared sense of wonder as we seek to uncover the mysteries of the cosmos and our place within it. In doing so, we embrace the fundamental truth that the universe is not just a backdrop for our story but a vast expanse brimming with potential, waiting for us to explore and understand.

Miles Kepler

# Chapter 2: Conditions for Life: The Goldilocks Zone

When we think about the possibility of life beyond our beautiful blue planet, we first need to explore what makes Earth such a perfect home for living things. The idea of "habitability" might seem simple, but it includes a fascinating mix of environmental and planetary conditions that together create a nurturing space for life. Imagine a delicate balance of factors where temperature, atmosphere, gravity, and the presence of certain chemicals all work together to support the growth and survival of living organisms.

At the core of our understanding of habitability are some key criteria that help us set the boundaries. Temperature is especially important; if it gets too hot, water will evaporate, leaving any potential life thirsty and lifeless. If it gets too cold, everything can freeze over, creating a barren wasteland where life can't thrive. The ideal temperatures for liquid water, which is essential for life, fall within a cozy range scientists call the "Goldilocks Zone." This zone is neither too close to a star, where everything would be scorched, nor too far away, where it would turn into a frozen landscape. Instead, it's that sweet spot where conditions are just right for life.

As we think about the Goldilocks Zone, we also need to consider a planet's atmosphere. The atmosphere acts like a protective blanket, keeping harmful radiation at bay while also ensuring the pressure is just right for water to stay liquid. A planet with a thin atmosphere might face extreme temperature swings, much like a desert where scorching sun during the day gives way to icy cold at night. On the flip side, a thick atmosphere can trap heat, causing a runaway greenhouse effect that turns the surface into an unlivable oven. The right mix of gases, particularly oxygen and nitrogen, can mean the difference between a barren rock and a lively ecosystem bustling with life.

Gravity is another key player that often goes unnoticed in the conversation about habitability. A planet's gravity must be strong enough to hold onto its atmosphere and liquid water but not so heavy that it crushes any potential life forms. Earth's gravity strikes the perfect balance, allowing it to keep an atmosphere while giving living organisms the lift they need to thrive. Picture a planet where gravity is too weak, causing its atmosphere to drift away like a helium balloon floating into the sky. In such a case, any chance for life would vanish.

While our understanding of habitability largely comes from conditions found on Earth, we must also appreciate the incredible adaptability of life. The rich diversity of life here, from the lush rainforests to the stark arctic tundras, broadens our understanding of what a habitable environment might look like. Take extremophiles, for example—those amazing organisms that flourish in places we once thought were impossible for life. Hydrothermal vents deep in the ocean are alive with creatures despite the intense heat and darkness. These vents spew boiling water full of minerals, creating ecosystems that thrive without any sunlight. If life can survive in such extreme conditions, who's to say it couldn't also adapt to other environments out in the universe, possibly on far-off planets or moons?

Exploring these extreme environments on Earth shows us a larger truth: life might not be as choosy as we once believed. Acidic lakes, frozen tundras, and super salty environments all host thriving communities of organisms. These discoveries challenge the idea that life can only exist in Earth-like conditions, encouraging us to imagine what other habitable environments might be out there in the universe.

Looking up at the stars and the many exoplanets we've found, we can't help but wonder if other worlds also exist in their own Goldilocks Zones. The Kepler and TESS missions have uncovered thousands of exoplanets, and many of them are in areas that could potentially support life. Each new finding is like a breadcrumb, leading us along the path to discovering where life might exist beyond our own planet.

These missions have given us valuable insights into the variety of planetary systems out there. Some exoplanets are in their stars' habitable zones, where temperatures could allow for liquid water. However, others may be outside of this zone. Yet, as we've learned from extremophiles here on Earth, just having liquid water isn't the only thing that matters when it comes to habitability. Thanks to new technologies and innovative exploration methods, we're refining our understanding of what makes a planet livable. We're becoming more open to the idea that life could thrive in ways we haven't even thought of yet.

As we dive deeper into this exciting realm of possibilities, we also need to pay attention to the chemical elements that are the building blocks of life. Elements like carbon, nitrogen, oxygen, and phosphorus come together to form the complicated molecules necessary for living organisms. Carbon, in particular, is incredibly versatile, creating a wide range of compounds that are essential for life, like carbohydrates, proteins, lipids, and nucleic acids. Understanding where these elements are found in the universe gives us clues about where life might arise.

While we explore these amazing possibilities, we can't overlook the crucial role of water. It's a vital part of nearly every biological process we know. Often called the solvent of life, water helps drive essential chemical reactions and acts as a transport system for nutrients and waste. Just finding water is a promising sign that life could exist in some form, even if it doesn't look like the life we're familiar with.

Scientists are now looking for liquid water on the icy moons in our solar system, like Europa and Enceladus, where hidden oceans lie beneath thick layers of ice. Even though these moons are far from the sun's warmth, they might have conditions suitable for life, thanks to geothermal activity from their planetary bodies. If we're going to find extraterrestrial life, we need to explore not just rocky planets but also the cold, icy worlds that are often overlooked.

As we think about what defines habitability, we can't ignore the deep implications of our findings. Exploring other worlds and searching for life beyond Earth isn't just a scientific mission; it taps into our natural curiosity about our place in the universe. Every new discovery, whether it's a fresh exoplanet or an extremophile thriving in a harsh environment, reminds us of how diverse and resilient life can be.

The Goldilocks Zone serves as a tantalizing guide in our search for extraterrestrial life, encouraging us to consider the many possibilities that might exist beyond the stars. It challenges us to expand our ideas of what makes a place livable, pushing the boundaries of our imagination. As we keep exploring the universe, we must stay open to unexpected surprises, for it is in the uncharted territories of existence that we might uncover the answers to our biggest questions.

**Discoveries from Kepler and TESS Missions**

In the vastness of space, where stars twinkle like fireflies against the dark sky, the search for planets that might host life has taken on a new sense of excitement and urgency. NASA's Kepler and TESS missions have pushed our exploration of the universe to new heights, revealing a cosmos full of wonders that invite us to dig deeper. These missions have transformed our understanding of exoplanets—those distant worlds orbiting stars beyond our solar system—and offered us a glimpse of environments that might just be suitable for life.

Picture yourself lying on a grassy hill under a starry sky, the Milky Way arcing overhead like a celestial river. As you look at the shimmering lights, each representing a distant sun, you can't help but wonder about planets similar to ours, quietly spinning in the vastness. This sense of wonder lies at the heart of what the Kepler and TESS missions strive to uncover—the potential for life on other worlds.

Launched in 2009, the Kepler Space Telescope was a groundbreaking project with a clear goal: to find Earth-sized exoplanets in the habitable zones of their stars. It captured the imagination of scientists and sky gazers alike, becoming a beacon of hope in our search for life beyond Earth. Kepler used a clever technique called the transit method, which involves watching the brightness of stars. When a planet passes in front of its star, it briefly blocks some of the light, creating a tiny but noticeable dip in brightness. This method allowed Kepler to detect planets that might otherwise remain hidden in the cosmos.

During its nine-year mission, Kepler uncovered more than 2,600 exoplanets—a staggering achievement that reshaped our understanding of the universe. Each new planet discovery felt like finding a missing piece of a grand puzzle, shedding light on how planets form and evolve. One of the most exciting revelations from Kepler was the realization that Earth-like planets are much more common than we once thought. The data suggested there could be billions of such planets in our Milky Way galaxy, cozy in their stars' Goldilocks Zones—the perfect spots where temperatures are just right for liquid water.

The enthusiasm surrounding Kepler's discoveries was electric. Each announcement of a new exoplanet felt like unearthing hidden treasure. A standout among these was Kepler-186f, the first Earth-sized planet found in the habitable zone of another star. This discovery sent shockwaves through the scientific community and sparked imaginations everywhere, inspiring dreams about what life might be like on this distant world. Kepler-186f orbits a star that's cooler and dimmer than our sun, but its location in the habitable zone hints at conditions that could support life. It's a reminder of our place in the universe and the astonishing possibilities waiting just beyond our reach.

As the Kepler mission wrapped up, it passed the baton to the Transiting Exoplanet

Survey Satellite, or TESS, which launched in 2018 with big ambitions. TESS was built to extend Kepler's legacy by surveying a larger swath of the sky and focusing on stars that are closer to us. This is crucial because the nearer a star is, the easier it is to study its planets in depth. TESS also uses the transit method but with a wider field of view, allowing it to scan an impressive 85% of the sky.

In its short time of operation, TESS has already made remarkable discoveries. In its first year alone, it identified thousands of candidate exoplanets, showcasing the mission's efficiency and effectiveness. TESS doesn't just look for new planets; it also aims to understand these worlds better to figure out their makeup and atmospheres. These insights are vital because they help us assess whether these planets might host life.

One of TESS's most thrilling discoveries is TOI-700 d, an Earth-sized exoplanet in the habitable zone of its star. Just 100 light-years away, this planet presents an exciting opportunity for further research. With its size and proximity, TOI-700 d has piqued the interest of scientists eager to learn more about its atmosphere and potential for supporting life. Plans are already in place to observe this planet with powerful telescopes, allowing researchers to gather more data and perhaps one day answer the exciting question: Is there life on TOI-700 d?

The search for habitable worlds isn't just about numbers; it also invites us to imagine what life might look like on these distant planets. While we often use Earth as our reference for habitability, the discoveries from Kepler and TESS encourage us to think creatively about life in different settings. Just as there are extremophiles on Earth that thrive in harsh conditions, it's reasonable to think that life on exoplanets could take forms we can hardly imagine.

Think about the implications of finding a planet in the habitable zone around a red dwarf star, which is cooler and dimmer than our sun. These planets might be tidally locked, meaning one side always faces the star while the other remains in constant darkness. The varying conditions on these planets could give rise to unique ecosystems, with life forms adapting to the stark differences between light and shadow. This kind of imaginative thinking fuels the passion of astronomers and scientists as they continue to explore the data collected by Kepler and TESS.

Moreover, the information gathered by these missions has illuminated how frequent Earth-like planets are in our galaxy. The more we understand about planetary systems, the clearer it becomes that our solar system is just one of many. Each new discovery pushes us to rethink our assumptions and broaden our definitions of what makes a planet habitable. We now realize a planet doesn't have to be Earth-sized or orbit a sun-like star to support life; even smaller, rocky worlds around red dwarf stars may hold the secrets to life as we know it.

As we reflect on these discoveries, we can't overlook the role of water, the essential ingredient for life as we know it. Many of the exoplanets found so far lie in regions where liquid water could exist, either on the surface or underneath thick ice layers. This realization has refreshed our perspective on where life might arise, even in environments vastly different from our own.

Searching for water on exoplanets is like being a treasure hunter, sifting through the sands of the universe hoping to find the next big discovery. Thanks to advancements in telescopes and detection techniques, scientists can now analyze the atmospheres of these planets for signs of water vapor and other crucial indicators of habitability. The possibility of finding water on otherwise inhospitable planets is an exhilarating prospect that fuels the imaginations of scientists and aspiring explorers alike.

Take, for example, the icy moons of our solar system, like Europa and Enceladus, which have been recognized as promising spots for life. These moons, with their subsurface oceans, remind us that life can thrive in unexpected places. As we broaden our search beyond just rocky planets, the lessons learned from Kepler and TESS inspire us to explore every corner of the solar system and beyond, expanding our definition of what makes a world habitable.

The importance of the Kepler and TESS missions goes well beyond the discoveries they have made; they have reshaped how we view the cosmos and our role in it. Each new exoplanet found represents a fresh possibility, a new hope that life might exist beyond our planet. This quest for knowledge not only has scientific implications but also raises philosophical questions. As we learn more about the universe around us, we find ourselves grappling with our own existence and the potential for life beyond Earth.

These missions have ignited a renewed sense of curiosity, inspiring a generation of scientists, astronomers, and dreamers to look up at the stars. They remind us that we are not merely spectators of the universe; we play an active role in the wonderful story of discovery. Each planet discovered by Kepler and TESS brings us one step closer to answering the timeless question: Are we alone?

Ultimately, the Kepler and TESS missions are more than just achievements in technology; they symbolize humanity's unyielding pursuit of knowledge. They embody the spirit of exploration and the desire to understand our universe. As we stand on the shoulders of those who came before us, gazing at the stars, we are filled with hope and wonder, knowing that the next great discovery could be just around the corner—or light-years away. The universe, in all its vastness, continues to inspire us to dream, to explore, and to seek answers to the profound questions that shape our understanding of existence.

**The Role of Water in Life's Existence**

Water, that simple yet amazing substance, is something we often overlook in our everyday lives. We splash it, drink it, bathe in it, and let it flow endlessly, hardly ever pausing to think about how truly important it is. But when we consider life beyond our blue planet, water becomes vital in ways we can't ignore. The role of water in supporting life isn't just about biology; it's a key part of existence itself.

At its essence, water is the perfect foundation for life as we know it. It acts as a solvent, a delivery system, a temperature stabilizer, and it plays a part in countless chemical reactions that keep life going. Picture the biochemical processes happening inside a cell, as enzymes and molecules interact in a beautiful dance. Within this watery realm, the magic of life happens, where organic molecules come together to form the building blocks of living things. Without water, the complex structures that make up proteins, nucleic acids, and carbohydrates would simply fall apart—like salt dissolving in dry desert air.

So, what makes water so special for life? To start, its molecular structure may seem simple at first glance, but it has a huge impact. Water consists of two hydrogen atoms bonded to one oxygen atom ($H_2O$), and its bent shape creates a polar nature. This polarity allows water molecules to cling to each other through hydrogen bonds, leading to some remarkable properties. The way water sticks to itself and to other surfaces helps it travel through plants from roots to leaves—a vital process for their survival. Plus, water has a high specific heat capacity, which means it can absorb and release heat without making temperatures swing wildly, creating stable environments where organisms can thrive. And how could we forget that unique quirk where water expands when it freezes, allowing ice to float? This is crucial for the survival of many aquatic ecosystems.

When we think about the significance of water here on Earth, it's natural to wonder what it means for life on other planets. The hunt for liquid water on other celestial bodies has become a major focus in astrobiology. We've explored our solar system, searching for signs of water in every form. From the icy caps on Mars to the geysers shooting vapor from the moons of Jupiter and Saturn, the evidence of water is both exciting and thought-provoking. Could these places hold tiny life forms, or even life forms we can't even imagine?

Let's start with Mars, our closest neighbor. The mystery of the Red Planet has fascinated people for ages, but it's the discoveries made in recent years that have really sparked our curiosity. We've found evidence of ancient riverbeds, lakebeds, and minerals that only form with water, suggesting that Mars was once a much wetter place. Rovers equipped with advanced technology have discovered signs of salty water and dark streaks on its slopes, which might mean liquid water is just beneath the surface. Scientists think that these conditions, even if only temporary, could have created an environment friendly to microbial life. The idea that life might have once thrived on

Mars—or could still exist there in some form today—is an exciting possibility that highlights the importance of water in our search for life beyond Earth.

Meanwhile, the icy moons of Jupiter and Saturn have their own secrets to reveal. Europa, one of Jupiter's largest moons, appears frozen over with a thick layer of ice. But beneath that chilly exterior lies a vast ocean, possibly holding twice the amount of all of Earth's oceans combined. The thought of a hidden ocean filled with life makes Europa one of the most promising places to search for extraterrestrial organisms. Scientists are considering missions to send landers or even probes below its icy crust to look for signs of life. If life is out there, it might be completely different from what we know, adapting to the dark and high-pressure conditions beneath the ice, thriving in a world where sunlight can't reach.

Similarly, Enceladus, one of Saturn's moons, has caught our attention with its active geysers that shoot water vapor and organic molecules into space. These geysers, caused by the gravitational pull of Saturn, provide an exciting way to study the moon's hidden ocean without landing a probe. The Cassini spacecraft, which explored Saturn and its moons from 2004 to 2017, found complex organic compounds in these plumes. With liquid water, organic molecules, and energy sources all present, Enceladus is a top contender when it comes to the search for life. Discovering these geysers and examining what they contain has reignited our hopes for finding life beyond Earth, showing us that we need to think broader about where and how life might exist.

The search for water doesn't stop at our solar system; it stretches into the realm of exoplanets, where scientists are investigating the atmospheres of planets far away. Finding water vapor in the atmospheres of these exoplanets has opened up new ways to understand if they could support life. With powerful telescopes like the Hubble Space Telescope and the newer James Webb Space Telescope, researchers are looking at these distant planets' spectral signatures. They're on the lookout for clues of water vapor, which could suggest that these planets are in the habitable zone of their stars and have the right conditions for life.

Consider the recent discovery of exoplanets like K2-18b, a super-Earth about 124 light-years away that may have an atmosphere rich in water vapor. When scientists analyzed the data and found water, excitement surged among the research community. This finding highlighted the possibility that the universe might be filled with planets where liquid water exists, creating environments perfect for life. The implications of these discoveries reach far beyond our own Milky Way, challenging our understanding of what conditions are necessary for life to develop and thrive.

The importance of water in the hunt for extraterrestrial life is more than just a scientific curiosity; it taps into deep questions about our existence and place in the universe. As we consider the possibility of life on other worlds, we can't help but think about our own beginnings. Earth's story is closely tied to water, from the early oceans where the first simple organisms appeared to the lush ecosystems that now thrive all over our planet. Water has shaped our Earth, influenced our evolution, and ultimately provided the cradle where life began.

This connection between water and life leads us to wonder about what might lie beyond our planet. Could there be other forms of life, perhaps vastly different from what we know, living in the oceans beneath the frozen surfaces of distant moons? Could there be organisms that rely on different solvents or biochemistries, thriving in environments we can hardly imagine? These inquiries fuel our scientific curiosity, pushing us to explore the universe and broaden our understanding of life's potential.

As we stand on the brink of new discoveries, a sense of wonder surrounds us. The quest for water on distant worlds isn't just about finding a chemical compound; it's an exploration of life itself, a journey into what it means to exist among the stars. Each new finding sparks our imagination and drives us further into the unknown, reminding us that the universe is filled with endless possibilities just waiting to be uncovered.

With every mission, discovery, and data point collected, we get closer to answering the timeless question: Are we alone? Water's role in our search for life goes far beyond Earth; it invites us to look beyond, to dream, and to seek a deeper understanding of the universe surrounding us. Water, in its simplicity, remains a powerful symbol of hope and curiosity, guiding our quest for answers in the vast night sky. So let's embrace this adventure with open hearts and minds, for the story of water and life is far from over. The pages of our cosmic narrative are still being written, and the ink flows freely through the currents of time and space, waiting for us to explore its depths.

Miles Kepler

# Chapter 3: Unraveling Mars: Close to Home

When you look up at the night sky, the red planet that twinkles back at you isn't just a distant speck among countless stars; it symbolizes our closest connection to the universe. Mars, often lovingly called Earth's celestial twin, has captivated the imaginations of astronomers, scientists, and stargazers for centuries. Its striking rust-like color hints at a mysterious past hidden beneath its dusty plains and towering volcanoes. The allure of Martian mysteries pulls at the hearts of dreamers, leading us to wonder: could this nearby world have once supported life?

To truly grasp the bond between Earth and Mars, we should first explore their similarities in geology and atmosphere. The two planets share several key traits that make Mars an exciting target for exploration. For instance, both planets have a day length that is surprisingly similar. A single day on Mars—known as a sol—lasts about 24 hours and 37 minutes. This slight difference helps scientists imagine the rhythms of life that could have existed on Mars, echoing those thriving here on Earth. It's almost as if these planets are sharing secrets across the vastness of space.

In addition to having similar day lengths, Mars experiences seasons just like Earth. Its axial tilt of around 25 degrees creates various climates and weather patterns. This tilt results in seasonal changes that dramatically alter the Martian landscape throughout the year. Polar ice caps, mostly made of water and dry ice, expand and shrink with the seasons, creating a captivating show that hints at the planet's complex climate. The surface of Mars is marked with ancient riverbeds and channels, suggesting that liquid water once flowed freely, raising exciting questions about the possibility of past life.

Mars also boasts its own mountains, valleys, and even the largest volcano in the solar system—Olympus Mons. Towering at an astonishing 13.6 miles high, this massive shield volcano is nearly three times taller than Mount Everest. Its immense size and gentle slopes indicate it formed over millions of years from lava flows, just like volcanoes on Earth. Meanwhile, the expansive canyon system known as Valles Marineris stretches over 2,500 miles long and plunges up to seven miles deep, offering a geological wonder that challenges our understanding of how planets evolve. The sheer scale of these features invites us to imagine the forces that shaped

Mars, revealing a history that may not be so different from our own planet's.

However, for all these similarities, Mars is a world that is very different from Earth. Its atmosphere is thin compared to the thick layer that protects our planet. Made mostly of carbon dioxide with only tiny amounts of oxygen, it is unwelcoming to most forms of life as we know it. The surface pressure on Mars is less than one percent of that on Earth, creating conditions where water can't stay in liquid form for long. The harsh reality includes frigid temperatures that can drop to minus 195 degrees Fahrenheit at the poles, painting a stark picture of a planet that feels both familiar and foreign.

Yet, this contrast is what truly captivates our imagination. The quest to uncover Mars's secrets isn't just a scientific mission; it's a deep journey into what it means to be human. The story of Mars is woven into our own, prompting us to reflect on our existence and our place in the universe. The possibility of microbial life on Mars, however tiny, makes us ponder our origins and the essence of life itself. What does it mean to search for life beyond Earth? Is it a quest for knowledge, or a deeper longing for connection in this vast cosmos?

The tale of Mars is also a testament to human creativity and determination. Space missions have brought us closer to understanding our planetary neighbor, revealing the Martian landscape in vivid detail. The Mariner spacecraft first captured images of the Martian surface in the 1960s, sparking a wave of curiosity that has since led to significant exploration efforts. NASA's Viking landers came next, probing the surface for signs of microbial life in the 1970s. Although they didn't find clear evidence of life, they laid the groundwork for future explorations and opened up a world of questions.

    Fast forward to the 21st century, and the excitement of Mars exploration is still alive and well. Rovers like Spirit, Opportunity, and Curiosity have roamed the planet's surface, sending back stunning images and valuable data about its geology and climate. Each rover has expanded our understanding of Mars, uncovering minerals that form in the presence of water and identifying ancient environments that might have been suitable for life. These discoveries act as breadcrumbs on a path of curiosity, encouraging us to dig deeper and explore further into Mars's past.

    The most recent rover, Perseverance, launched in July 2020, has taken this journey to a whole new level. Equipped with cutting-edge instruments, it aims to find signs of ancient life and gather samples that could one day return to Earth. NASA's bold plan to send humans to Mars in the coming decades speaks to our relentless curiosity and desire to explore the unknown. The vision of humans setting foot on Martian soil is more than just a dream; it inspires generations of scientists, engineers, and dreamers, all of whom stand on the brink of a new frontier.

As we study the red planet, we are reminded of our place in the universe. Mars tells a story of resilience and adaptation, a tale that mirrors our own. The search for extraterrestrial life isn't just about finding living beings in far-off places; it's about exploring what life means, how it thrives, and the connections that unite us all. The quest to unlock Mars's secrets reflects our shared human experience—a journey fueled by wonder and an unquenchable thirst to understand our existence.

In recent times, discussions about Mars have shifted from exploration to colonization. As we face challenges like climate change and dwindling resources on Earth, the idea of creating a human presence on Mars becomes more appealing. The concept of terraforming, or changing the Martian environment to make it more like our own, excites the minds of scientists and visionaries alike. What if we could turn Mars into a second home, a safe haven for humanity?

While such dreams might sound ambitious, the idea of building a self-sustaining colony on Mars has been taken seriously in research circles. Ideas for habitats, life support systems, and resource utilization are beginning to take shape, driven by what we've learned from past missions. The prospect of using Martian resources—like extracting water from the polar ice caps and using local materials for construction—offers a fascinating glimpse into what the future might hold.

However, as we dream of colonizing Mars, we must proceed with caution. The ethical questions surrounding the alteration of another planet present tough dilemmas. Should we reshape a world that may still hold undiscovered secrets? Our pursuit of knowledge and exploration should not come at the cost of preserving the integrity of other celestial bodies. Mars is more than just a possible second home; it is a historical site, a testament to the evolution of our solar system, and potentially, a birthplace of life.

As we gaze into the future and think about our relationship with the red planet, we recognize that the journey is just as important as the destination. Each rover traversing the Martian landscape, each satellite surveying its surface, contributes to a larger story that connects us to something beyond ourselves. This adventure of discovery cultivates a sense of unity, bridging divides between cultures and inspiring collaboration among nations.

In a world that often feels fractured, the search for extraterrestrial life has the power to bring us together in a common mission—one that transcends borders and reflects our shared desire for understanding. The exploration of Mars encourages us to look beyond our immediate surroundings, inviting us to embrace the vastness of the universe and our place within it.

Mars, our celestial twin, is not merely a neighboring planet; it mirrors our own world, acting as a canvas for our hopes and dreams. As we work to uncover the secrets of the red planet, we are, in a way, unraveling the mysteries of ourselves. The search for life beyond Earth is, fundamentally, a search for connection, purpose, and the reassurance that we are not alone in this endless universe. With each new discovery, we move closer to not just understanding Mars, but also grasping the profound implications of our existence in the cosmos.

### The Legacy of Mars Missions

The history of Mars exploration is more than just a timeline of scientific breakthroughs; it's a colorful story filled with human curiosity, dreams, and the unyielding quest for understanding. As we take a closer look at this journey, we'll uncover the captivating narrative of our efforts to learn about the red planet—a tale dotted with victories and setbacks, debates and discoveries. The mystery of Mars has sparked imaginations and inspired countless scientists, dreamers, and everyday people who gaze up at the stars.

Interest in Mars started long before we sent our spacecraft into the cosmos. Back in the 19th century, astronomers with the best telescopes of their time peered into the universe, captivated by the bright red point of light that stood out against the night sky. One prominent figure, Italian astronomer Giovanni Schiaparelli, reported seeing what he thought were "canali," or channels, on the surface of Mars. Unfortunately, this was misinterpreted as "canals" in English, igniting wild speculation about the possibility of advanced civilizations on Mars. The idea that intelligent beings could exist on our neighboring planet captivated the public and spurred a wave of creative works in literature and film, all imagining life on Mars.

While Schiaparelli's observations were groundbreaking, they were also tinted with the romantic notions of the time. What we later discovered about Mars was far more intricate. The excitement surrounding the possibility of a Martian civilization paved the way for a century of exploration. This enthusiasm initiated a series of missions aimed at uncovering the secrets of Mars, setting the stage for future endeavors.

Fast forward to the 1970s when the space age finally made the dreams of those early astronomers a reality. Enter the Viking missions: Viking 1 and Viking 2, launched by NASA, which were the first to land on Martian soil and carry out experiments right there. These landers came equipped with advanced tools designed to not only capture stunning photos of the landscape but also analyze the soil and look for signs of life. They conducted a variety of tests to check for microbial life, and the results sparked heated debates among scientists.

The Viking landers sent back an impressive array of images showcasing the different features of Mars. The stark, rocky plains and massive canyons looked nothing like the bustling civilization some had imagined. The most controversial aspect of the Viking missions turned out to be the life detection tests. The results were unclear, leading to a split among scientists—some believed the data hinted at life, while others argued it was just a chemical response. This confusion set the stage for future missions to keep searching for life, fueled by the excitement of earlier successes and the lessons learned from their uncertainties.

The impact of the Viking missions was significant, as they laid a strong foundation for future explorations. They helped scientists refine their research methods and rethink their ideas about what Mars might hold. The knowledge gained from the Viking missions influenced the design of upcoming missions and deepened our understanding of the planet. The excitement that had started with the first telescopic views of Mars grew into solid data collected straight from its surface.

As we entered the 21st century, a new chapter of exploration began with the introduction of rovers designed to roam Mars. The Spirit and Opportunity rovers, launched in 2003, took exploration to a whole new level, delving into the Martian landscape in ways stationary landers couldn't. These rovers were more than just machines; they were symbols of human curiosity. Spirit settled in Gusev Crater, while Opportunity landed in Meridiani Planum, sending back breathtaking images of the Martian terrain and revealing signs of past water.

By moving across various geological features, both rovers greatly expanded our scientific knowledge. Opportunity, in particular, far surpassed expectations, functioning for nearly 15 years instead of the planned 90 days. Its journey across the rusty terrain uncovered remarkable findings that suggested Mars may have once had conditions fit for life. The discovery of hematite, a mineral typically formed in water, along with the detection of small spherical "blueberries," hinted at a wet past. These breakthroughs not only sparked scientific curiosity but also reignited the public's interest in Mars.

The legacy of these rovers lies not just in their scientific accomplishments but also in their tales of resilience. The engineers and scientists behind these missions faced numerous challenges, from unexpected equipment malfunctions to the harsh Martian climate. Each challenge they overcame showcased human creativity and teamwork. The rovers became symbols of hope and determination, reflecting the spirit of exploration that drives us to seek answers beyond our planet.

Continuing this path of exploration, the Curiosity rover, launched in 2011, marked a significant milestone in our Mars journey. With a range of advanced scientific tools, Curiosity was built to evaluate the planet's ability to support life. Its travels through Gale Crater quickly unveiled a treasure trove of geological history, revealing ancient riverbeds and organic molecules—essential components for life. These discoveries elevated the search for past life and ignited discussions about the potential for life on Mars today.

Curiosity's ability to analyze rock samples and gather data in real-time captured the imagination of people around the globe. Its stunning images of Martian landscapes, showcasing rugged mountains and wide-open plains, inspired a sense of adventure and discovery. Every new finding brought a wave of excitement—a collective thrill that resonated with scientists and the public alike. The team behind Curiosity became a close-knit community, fueled by their passion for exploration and their drive to push the boundaries of knowledge.

The legacy of Mars missions extends beyond just technology and scientific discoveries; it's also about the human stories that weave through these explorations. The engineers, scientists, and technicians who pour their hearts into these missions have personal connections to their outcomes. Their passion fuels their creativity, innovation, and ability to overcome obstacles. Behind every successful mission is a dedicated team that has spent years refining their skills, often working long hours, driven by coffee and a shared vision.

The story of Perseverance, the latest rover launched in July 2020, embodies this spirit of collaboration and creativity. Perseverance isn't just a technological marvel; it represents years of research and development. Its mission goes beyond exploration; it aims to prepare for future human trips to Mars. With cutting-edge tools for collecting and analyzing samples, Perseverance is on the hunt for signs of ancient life and gathering samples to eventually return to Earth. The rover's clever design mirrors the hope it represents—that one day, we may walk on the surface of Mars.

    Each mission to Mars is a stepping stone, building on the successes and setbacks of those that came before it. With every new discovery, we gain valuable insights into the planet's past, climate, and potential for life today. But more than that, these missions spark a sense of wonder and curiosity that transcends scientific inquiry. They encourage us to reflect on our own existence, our place in the universe, and the endless quest for knowledge that defines us as humans.

    The legacy of Mars exploration is also interwoven with culture. Books, movies, and art have long found inspiration in the red planet, mirroring our dreams and fears about our cosmic neighbor. As we ponder the implications of our explorations, we find ourselves in conversation with both the past and the future—wondering not only what we might uncover on Mars but also what it means for our understanding of life as a whole.

As we gaze up at the night sky, we remember that Mars is more than just a twinkling dot; it's a realm of possibilities. The legacy of Mars missions continues to deepen our understanding of life beyond Earth, pushing us to explore the far reaches of our solar system and beyond. Each mission stands as a testament to human creativity and our shared desire to solve the mysteries of the universe.

In this age of rapid technological progress and an increasing urgency to explore new frontiers, the legacy of Mars missions shines as a beacon of hope, guiding us toward future explorations. The drive to reach for the stars isn't just about science; it's deeply rooted in our human nature—the innate urge to explore, discover, and connect with something greater than ourselves. As we forge ahead in our exploration of Mars, and potentially land there as humans one day, we carry with us the stories, discoveries, and aspirations of those who came before us. The spirit of exploration is very much alive, and with every new mission, we are reminded that the journey to understand our universe has only just begun.

## Investigating Microbial Potential

The hunt for microbial life on Mars has fascinated scientists, adventurers, and dreamers for years. For decades, Mars has been seen not just as a mysterious neighbor in our solar system, but also as a possible home for life beyond Earth. Imagine if, hidden beneath its dusty surface and thin atmosphere, tiny alien microbes are waiting to be found! The exciting possibility of life on Mars has encouraged researchers to design and carry out more advanced experiments, each aiming to answer questions that have puzzled us for centuries.

This quest for microbial clues began back in the 1970s with the Viking missions. These landmark missions marked humanity's first serious effort to land on Mars and look for signs of life. The Viking landers were equipped with a variety of scientific tools carefully crafted to analyze soil samples and conduct experiments to detect life. However, the results were confusing. The landers painted a picture of a Martian landscape that seemed barren and lifeless, but the lack of clear answers only sparked more curiosity.

The Viking missions created a divide among scientists. Some believed the mysterious signals picked up by the landers could indicate some form of microbial life, while others were cautious, urging against jumping to conclusions. This uncertainty fueled a deeper exploration of the topic, sparking a revival in astrobiology. Scientists began to think more creatively about what life could look like and where it might survive—not just in Earth's lush environments, but in some of the harshest settings imaginable.

Now, we find ourselves in a new age of exploration. Missions like the Mars 2020 Perseverance rover are changing how we search for life on the Red Planet. Launched with high hopes and cutting-edge technology, Perseverance signifies a major step forward in understanding what makes a planet capable of supporting life. As it roams the Martian landscape, the rover uses advanced instruments to gather rock and soil samples that will eventually be brought back to Earth for detailed study. Each sample collected holds the promise of unlocking new insights into the planet's history, hinting at whether life ever existed beneath its surface.

One of the key parts of Perseverance's mission is its search for biosignatures—these are molecules, patterns, or markers that hint at biological activity. This could range from organic molecules to specific isotopic ratios that suggest life processes. As the rover explores the rocky terrain, excitement builds;

every rock and every bit of soil could reveal remnants of ancient life.

When we think about the possibility of life on Mars, the idea of extremophiles becomes very relevant. On Earth, extremophiles are remarkable organisms that thrive in conditions once thought to be unlivable. You can find them in the extreme heat of deep-sea hydrothermal vents, the freezing temperatures of Antarctic ice, or the acidic waters of volcanic lakes. These tough organisms have broadened our understanding of where life can exist, and they are crucial for astrobiologists searching for similar life on other worlds.

The implications of extremophiles regarding Mars are significant. If life can survive in such extreme conditions on Earth, it makes it easier to believe that life could have developed on Mars when it was wetter and warmer. Perseverance's mission aims to investigate whether conditions once existed on Mars that might have supported similar life forms. The rover's tools are built to analyze the soil and atmosphere in search of organic compounds and other biosignatures that could point to ancient microbial life.

The thrill surrounding these discoveries is impossible to ignore. With each new piece of data, scientists are eager to weave together the story of Mars' geological and climatic past. Did ancient rivers shape the landscape, possibly supporting life? Did lakes once fill the planet's basins, creating ideal habitats for microbial life? As researchers analyze the samples collected by Perseverance, they are excited to uncover the hidden chapters of Mars' history.

Looking ahead, upcoming missions—like the much-anticipated sample return missions—will be vital to this ongoing investigation. These missions aim to bring Martian soil and rock samples back to Earth for even more in-depth analysis. This next phase of exploration not only holds the promise of groundbreaking discoveries but also raises deep questions about our place in the universe. What would it mean if we found evidence that life once flourished on Mars? How would that change our understanding of life on Earth and the possibility of it existing elsewhere?

These questions go beyond academic curiosity; they tap into our core desire to explore and understand. As we reach the edge of human exploration on Mars, the dream of finding microbial life shifts from mere possibility to a real potential. The idea of humans walking on Martian soil adds a whole new layer to this already fascinating story. With boots on the ground, scientists will have the chance to study the Martian environment directly, in ways we simply can't do from afar. This could lead to breakthroughs that have eluded us for years—or even centuries.

In the end, the search for microbial life on Mars brings together scientific rigor and human curiosity. It encourages us to question what life really is and how we connect to the universe. The quest to discover whether we share our universe with other living beings challenges how we see ourselves and fills us with an ever-growing sense of wonder.

In this fast-paced age of technology, the excitement surrounding Mars exploration is contagious, sparking public interest and scientific inquiry alike. As we invest time and creativity into our missions, we aren't just looking for microbial life; we're also exploring what it truly means to be human. We're united in this quest, driven by our shared thirst for knowledge and a connection to something larger than ourselves. Each discovery on Mars is a tribute to our collective imagination, a reminder of the power of exploration, and an invitation to dream even bigger.

By building on the lessons learned from past missions, utilizing cutting-edge technology, and drawing on the insights from extremophiles, we find ourselves at a crucial point in the search for life beyond our planet.

The excitement is palpable, the journey is profound, and the potential for discovery is vast. As we dig deeper into the microbial potential of Mars, we recognize that the exploration story is far from over. In fact, it's just beginning, with many pages yet to be filled with discoveries that could reshape our understanding of life in the universe. The mystery of Mars calls to us, and with each day that passes, we draw closer to revealing its secrets, standing on the brink of a new chapter in our search for life—a journey that may not only answer age-old questions but also inspire generations to come.

# Chapter 4: Moons of Mystery: Europa and Enceladus

The universe is an incredible and mysterious place, filled with countless celestial bodies that spark our curiosity. For centuries, people have looked up at the stars and wondered about what lies beyond our planet. Among these cosmic wonders, the icy moons of our solar system stand out as intriguing candidates for supporting life. Europa, a fascinating moon of Jupiter, and Enceladus, a captivating satellite of Saturn, are two prime examples in our ongoing search for extraterrestrial life. Both of these moons are believed to have subsurface oceans hidden beneath thick layers of ice, giving us a glimpse into environments that could possibly support life as we know it—or even forms of life we can't yet imagine.

Europa's icy surface is covered by a thick shell of frozen water, leading scientists to theorize that a vast ocean lies beneath. This ocean is thought to form due to the strong gravitational forces from Jupiter, which creates a unique environment where conditions for life might flourish. The gravitational pull generates heat through friction, preventing the water from freezing completely and allowing it to stay in a liquid state. As we imagine the icy world of Europa, we can't help but wonder about the possibility of life, both familiar and alien, lurking in its hidden depths.

Enceladus, on the other hand, makes a compelling case for being a home for life as well. This small but captivating moon, about the size of a city, has fascinated scientists and dreamers alike. Its icy surface is marked by geysers that shoot plumes of water vapor and organic molecules into space, hinting at a dynamic subsurface ocean that could potentially harbor life. In 2005, the Cassini spacecraft made an exciting discovery as it flew past Enceladus, revealing that these geysers were not just a quirky display of cryovolcanism, but a sign of a living, breathing world beneath the ice. This revelation sparked excitement among scientists, sparking debates and discussions about the possibility of life existing in this chilly environment.

Our understanding of what makes a place suitable for life has changed a lot over the years. We no longer limit our search for life to the warmer areas of planets; instead, we are exploring the icy reaches of our solar system, where extreme conditions—once thought to be uninhabitable—might actually provide the perfect environment for survival. A great example of this shift in perspective can be found in the hydrothermal vents on Earth. Located on the ocean floor, these vents are full of life that thrives in conditions of high pressure, extreme temperatures, and complete darkness. Creatures such as tube worms and extremophiles find energy in these harsh environments by relying on chemosynthesis instead of photosynthesis. These Earthly examples give us insight into what we might discover in the depths of Europa and Enceladus.

Scientists suggest that if hydrothermal vents exist on these icy moons, they could be perfect spots for microbial life to thrive. The combination of nutrient-rich water and chemical reactions in such places could create a rich mix of organic compounds, which are the building blocks of life. This idea encourages us to rethink where and how life might be found in the universe. It opens our minds to the possibility that life may not only be on planets with water on the surface but could also be flourishing underneath the protective ice that shields it from the harshness of space and cosmic radiation.

Imagining life forms in these extreme conditions ignites our curiosity. Could there be organisms that use chemosynthesis, much like those near Earth's hydrothermal vents, feeding off the chemical energy released from the moon's interior? Might we find unusual ecosystems with creatures adapted to the cold and dark, possibly glowing in the dark to attract mates or communicate? The potential for diverse life forms, shaped by the unique environments of Europa and Enceladus, taps into our desire to explore the unknown and challenges our understanding of what life really is.

While the possibility of finding life on these moons is thrilling, we also need to acknowledge the challenges ahead. The technology needed to investigate the icy layers of Europa and Enceladus has come a long way, but there are still hurdles to clear before we can unlock these celestial mysteries. The ice covering these moons is thick and formidable, sometimes several kilometers deep or even more. Any mission sent to explore must be designed to break through this icy crust and endure the harsh conditions that lie beneath.

Looking ahead, several upcoming missions are set to explore these captivating worlds. The Europa Clipper mission, which is planned for launch in the 2020s, aims to closely examine Europa's ice shell and subsurface ocean. With a variety of scientific tools on board, the spacecraft will gather information about the moon's surface, subsurface composition, magnetic field, and its potential for habitability. Meanwhile, the Enceladus Orbilander mission, proposed for the 2030s, aims to land on the moon's surface and analyze the geysers directly, collecting samples from the plumes to check for signs of life.

The excitement surrounding these missions fills us with anticipation because they symbolize humanity's relentless quest to understand our place in the universe. What might we discover beneath the icy shells of Europa and Enceladus? Will we find evidence of life, or perhaps come across environments that surprise us? As we prepare for these groundbreaking missions, we're reminded of the timeless questions that have fueled our exploration: Are we alone in this vast cosmos? Can life exist in forms and settings beyond our current understanding?

As we think about the icy moons of our solar system, we're prompted to reflect on the natural curiosity that drives our desire to explore. Searching for extraterrestrial life goes beyond just scientific research; it speaks to our fundamental need to understand the universe and our role in it. Europa and Enceladus serve as symbols of hope, reminding us that life may exist in the most unexpected places. They challenge us to rethink what it means for a place to be livable and lead us to consider that life, no matter how strange, might thrive beneath the frozen surfaces of these celestial bodies.

Ultimately, the icy moons of Europa and Enceladus are not just distant worlds; they reflect our own curiosity and thirst for discovery. As we gaze into the cosmos, we can't help but ask: What stories do these moons hold, and what lessons can we learn about the essence of life itself? Through exploration, we may not only uncover the secrets of these moons but also gain a deeper understanding of existence and our connection to the universe. As we set out into the unknown, we're driven by the age-old quest to answer one of the most profound questions of all: Are we truly alone in this vast, mysterious cosmos?

## Hydrothermal Vents: Life's Engines

Beneath the icy stretches of our oceans lies a vibrant world full of life that seems almost impossible to imagine. Hydrothermal vents, found along the ocean floor, are amazing ecosystems that thrive in places where light is scarce and temperatures reach extremes that would scare away most living things. Yet, in this seemingly harsh environment, life not only survives but flourishes in ways that stretch our understanding of what's possible. These underwater havens showcase nature's incredible ability to bounce back, proving that life can thrive even in the unlikeliest of places.

Hydrothermal vents act like underwater geysers, where superheated water bursting with minerals seeps from the Earth's crust, creating a special habitat for all sorts of creatures. When the hot water pours out of the vents and meets the cold ocean, it creates a striking contrast in temperatures that actually encourages a rich array of life. Surrounding these vents, you'll find sprawling communities of tube worms, giant clams, and countless tiny life forms known as extremophiles—organisms that have found a way to survive in conditions that would be deadly for most other living beings. These extremophiles are the hidden heroes of the deep sea, demonstrating nature's creativity and its knack for adapting to tough situations.

Take the tube worm, for example. This incredible creature, which can grow several feet long, lacks a mouth or digestive system. Instead, it forms a partnership with chemosynthetic bacteria that live inside its body. When the superheated, mineral-rich water flows out of the vent, it carries hydrogen sulfide, a toxic substance for most living things but a vital energy source for these bacteria. Through a process called chemosynthesis, the bacteria convert the hydrogen sulfide into organic compounds that the tube worms can use. This remarkable teamwork highlights how collaboration in nature can lead to amazing solutions.

Giant clams also illustrate how adaptable life can be in extreme conditions. Like tube worms, they have formed a symbiotic relationship with tiny, photosynthetic algae called zooxanthellae. These algae live within the clam's tissues and use sunlight to produce energy. Even though light is scarce in the deep ocean, these clams have found a way to make the most of the little light that does reach them by relying on the algae's ability to capture it. This is yet another example of how life can thrive, even when faced with seemingly impossible odds.

As we dive deeper into the lives of these unique organisms, we start to uncover the amazing biochemistry that helps them thrive in such tough conditions. Extremophiles have special proteins and enzymes that allow them to work well at high temperatures and pressures—conditions that would cause most other organisms' proteins to break down. For example, researchers have discovered enzymes from extremophiles that stay stable and functional at temperatures over 100 degrees Celsius. This extraordinary resilience opens the door to many exciting possibilities, from advancements in industry to breakthroughs in medicine. Studying these organisms not only broadens our understanding of biology but also inspires new ideas for solving complex challenges.

The existence of hydrothermal vents and the life they support makes us rethink where life can exist. They serve as a powerful reminder of the potential environments that might be found on icy moons like Europa and Enceladus, where similar conditions could allow life to exist. If creatures on Earth can thrive in the dark and extreme pressures of the ocean floor, could there be similar life on these distant moons? Scientists speculate that if hydrothermal vents exist beneath the icy surfaces, they might supply the energy and nutrients needed to sustain microbial life in those hidden seas. The idea of life on these moons becomes even more exciting when we consider that they may have similar geological processes as our own planet.

Discovering life in such surprising places changes how we view life's potential in the universe. For years, the search for alien life has been tied to finding water, sunlight, and stable climates. However, what we've learned from hydrothermal vents shows us that life can survive and even thrive in places once thought too harsh. This discovery expands the field of astrobiology, prompting researchers to look more widely across different celestial bodies for signs of life.

Reflecting on what we've learned from hydrothermal vents, we can't help but think more deeply about the nature of life itself. The creatures living in these extreme environments challenge us to rethink what we consider to be a "habitable" setting. Life is remarkably adaptable and resourceful; it evolves and thrives even in the harshest conditions. This raises an intriguing question: Where else in our universe might life be hiding?

The discovery of life thriving in the depths of our oceans sends our imagination soaring toward the icy worlds of Europa and Enceladus. What secrets might be waiting for us beneath their frozen surfaces? Could there be ecosystems similar to those found at hydrothermal vents, buzzing with life forms that challenge our current understanding? As we look ahead to future space explorations, we feel a strong sense of excitement. The idea of finding life on these moons goes beyond just scientific curiosity; it taps into our deep desire to understand our place in the universe.

Thinking about life in extreme conditions also leads us to reflect on our own resilience as humans. Throughout history, we have faced challenges, often emerging stronger and more creative when times get tough. This shared strength is mirrored in the organisms living around hydrothermal vents, reminding us of our connection to the larger story of life on Earth. Just as tube worms and giant clams have adapted to their unique environments, we too must be ready to adjust to whatever challenges we encounter—whether here on our planet or in our quest to explore the cosmos.

As we prepare for upcoming missions to Europa and Enceladus, we're reminded of how important curiosity and exploration are to being human. Every step we take toward uncovering the mysteries of these icy moons could lead to insights that change our understanding of life and existence. Discovering microbial life thriving in these moons would not only be a groundbreaking scientific accomplishment but would also challenge how we think about what it means to be alive.

The journey of exploration and discovery is a crucial part of who we are, pushing us to seek answers to the profound questions that have fascinated us for ages. Hydrothermal vents remind us that life can survive in extraordinary conditions, encouraging us to stay open to the incredible possibilities the universe might hold. As we brace ourselves for the unknown, we're not just driven by our quest for knowledge; we're fueled by the enduring hope that we are not alone in this vast universe.

The world of hydrothermal vents stands as a remarkable testament to life's resilience and adaptability, showcasing the incredible ways organisms survive in extreme environments. The discoveries made here reach far beyond our oceans, igniting curiosity and inspiring us to think about the potential for life on other worlds. The lessons learned from these underwater ecosystems challenge us to rethink our assumptions about where life can exist and motivate us to explore the hidden corners of our universe with fresh enthusiasm. As we set out on this journey of discovery, we realize that searching for life on other planets isn't just about finding new forms of existence; it's really about understanding the very essence of life itself and our place in the vast universe.

## Upcoming Missions to the Icy Worlds

As we look to the stars, the icy moons of Europa and Enceladus shine brightly as exciting places to explore. These far-off worlds spark our imagination with the possibility of finding life in some of the most extreme conditions imaginable. With each glance at the universe, we find ourselves on the cusp of a new era of remarkable missions aimed at uncovering the secrets hidden beneath their frozen surfaces. The thrill builds with the upcoming projects from both NASA and the European Space Agency (ESA). These missions go beyond simply searching for alien life; they play a crucial role in expanding our understanding of the solar system and our place in it.

Leading the charge is NASA's Europa Clipper, a spacecraft ready to set off on an incredible journey to Jupiter's icy moon, Europa. Set to launch in the next few years, Europa Clipper comes loaded with cutting-edge scientific instruments that will help it investigate Europa's thick ice shell and the ocean lurking below. Picture a fleet of advanced tools soaring through the chilly emptiness of space, each crafted to reveal the moon's hidden wonders. The mission's goals are ambitious and significant, as it aims to pinpoint possible habitats for life and explore how Europa's icy crust interacts with the mysterious ocean beneath.

One of the most thrilling parts of the Europa Clipper mission is its focus on the moon's ice shell. Scientists believe Europa is covered by a thick layer of ice, hiding a vast ocean of liquid water underneath. The science team is eager to learn how this ice shell connects with the subsurface ocean and what that means for the possibility of life. By using ice-penetrating radar and spectrometers, the spacecraft will study the surface composition, searching for biological markers that might hint at the presence of life. As the spacecraft glides over Europa's surface, it will gather data that could completely reshape how we think about life beyond Earth.

But there's more excitement on the horizon! The European Space Agency is also preparing its own adventure with the JUICE (JUpiter ICy moons Explorer) mission. Launching soon after NASA's Europa Clipper, JUICE will blaze a new trail in exploring not just Europa but also Ganymede and Callisto, two other icy moons of Jupiter. JUICE will dive into the unique environments of these moons, each with its own mysteries and potential for life. The mission aims to understand how the icy surfaces of these moons relate to their hidden oceans and whether those oceans could support life.

What makes JUICE especially groundbreaking is its collection of advanced technologies. The spacecraft will use ice-penetrating radar, a powerful tool for mapping the subsurface features of these moons. This technology will help scientists measure the thickness of the ice and find out if the ocean below touches the rocky mantle, a key ingredient for the potential for life. Additionally, JUICE's onboard spectrometers will analyze surface materials, looking for organic compounds and other clues that might indicate life or prebiotic chemistry. The mission plans to follow a path that includes multiple flybys, offering an incredible chance to gather extensive data from these celestial bodies.

The implications of these missions go far beyond just searching for alien life. They encourage us to think about the role icy worlds play in our overall understanding of the solar system. By studying these moons, scientists can gain insights into the processes that shape planetary systems, how celestial bodies form, and the potential for life to emerge in different environments. The knowledge gained from these missions could change how we define habitability and broaden our search for life to other celestial bodies that might have been overlooked.

As we await the outcomes of these missions, it's important to keep in mind the bigger questions they raise. What would it mean for humanity if we found life elsewhere in the universe? The prospect of such a discovery fills us with wonder and curiosity, making us reflect on our existence and our place in this vast cosmos. It prompts profound questions about life itself: Is life a rare occurrence, unique to Earth, or is it a common phenomenon spread throughout the universe? Each answer we uncover comes with implications that could change how we view life, consciousness, and our relationship with the cosmos.

The buzz surrounding these missions is contagious, but we must recognize that the quest for knowledge requires patience and determination. Space exploration is filled with challenges, and the journey to uncovering the secrets of these icy worlds will not be without its hurdles. Yet, it is precisely this journey that fuels our human spirit. Each challenge we overcome, each obstacle we face, brings us closer to understanding our universe.

In a world that often feels disconnected and isolated, the possibility of finding life beyond our planet unites us. It encourages us to think about the shared curiosity that connects us as humans. As nations come together on these groundbreaking missions, we create bonds that cross borders, sparking a collective desire for knowledge and exploration. The discoveries made on the icy worlds of Europa and Enceladus will not belong to any one nation; they will be a triumph for all humanity.

Moreover, the technologies developed for these space missions will have benefits that reach far beyond the study of life in space. The advancements in ice-penetrating radar and spectrometry could be valuable in various fields, such as climate science, geology, and medicine. The drive to understand the unknown promotes innovation and creativity, leading to solutions that can improve society.

As we stand on the threshold of these exploratory missions, a sense of urgency emerges. The universe calls us to stay curious and engaged. The quest to reveal the secrets of Europa and Enceladus is not just for scientists and researchers; it's an adventure we can all take part in. We can support scientific initiatives, advocate for funding and education in space exploration, and most importantly, keep our curiosity for the cosmos alive.

In today's world, accessing knowledge about space exploration is easier than ever. Public interest in astronomy and planetary science is thriving, fueled by social media, documentaries, and popular science books. Communities are coming together, embracing the excitement of space exploration and inspiring the next generation to dream big. The calls for a future where we actively seek to understand our universe resonate strongly, urging us not to avoid the unknown but to welcome it with open arms.

As we look ahead, let's remember the lessons learned from the resilience of life around hydrothermal vents. If life can thrive in the darkest and harshest environments on Earth, what possibilities lie waiting for us in the icy depths of Europa and Enceladus? The answers to these questions could reshape our understanding of life, reminding us that the universe is more vibrant and alive than we ever imagined. This sense of potential drives our exploration and inspires us to reach for the stars.

The missions to Europa and Enceladus represent more than a search for life; they embody humanity's thirst for knowledge and understanding. As we prepare to step into the unknown, let's carry with us a sense of curiosity that pushes us to explore not just the cosmos but the very essence of existence. The icy worlds await us, and with them comes the promise of a remarkable journey that could transform our understanding forever. The universe is vast, mysterious, and full of untold stories, and we are only just beginning to discover them. Let's embrace this adventure, as it invites us to uncover not only what lies beyond our planet but also the depths of our own humanity.

# Chapter 5: The Exoplanet Revolution

The sky has always sparked human curiosity. From the ancient Babylonians who mapped the stars to Greek philosophers pondering the mysteries of the universe, we have consistently looked upward, searching for answers to questions that have intrigued us for ages. The journey to find planets beyond our solar system has been nothing short of remarkable—a tale filled with creativity, determination, and a bit of luck. The first confirmed exoplanet discovery—a planet orbiting a star outside our solar system—marked a game-changing moment in how we understand our universe. It sparked a revolution in astronomy and opened the floodgates to the idea that we might not be alone in this vast cosmos.

In October 1995, Swiss astronomers Michel Mayor and Didier Queloz made a groundbreaking discovery: they found 51 Pegasi b, an exoplanet located in the constellation Pegasus. This gas giant, often called a "hot Jupiter" because of its close orbit to its star, was completely different from anything in our own solar system. This discovery sent shockwaves through the scientific community, leading to a surge in research and capturing the public's interest. Suddenly, the thought that Earth could be just one of many planets in an expansive universe felt like a real possibility.

However, the path to this discovery was not easy. For centuries, most people believed that our solar system was a unique oasis in a universe that contained no other planets. The idea of other worlds was often dismissed as mere science fiction. Yet, as we developed better observational tools and deepened our understanding of the cosmos, the question of whether there were other planets orbiting distant stars became more pressing.

The 20th century brought along a wave of technological advancements that set the stage for the hunt for exoplanets. The creation of more powerful telescopes allowed astronomers to look deeper into space with newfound clarity. At the same time, radio astronomy opened a whole new world, letting scientists detect celestial phenomena that had previously remained mysterious. Pioneers like Edwin Hubble expanded our understanding of the universe's enormity, laying the groundwork for future explorations of planets beyond our solar system.

As astronomy advanced, so did the methods for discovering exoplanets. One of the first and most effective techniques, known as the radial velocity method, measures the gravitational pull a planet has on its star. By observing the slight "wobble" in a star's motion, astronomers could infer the presence of an orbiting planet, even if it was invisible. Alongside this, transit photometry, which detects when a planet passes in front of a star and causes a brief dimming of its light, became essential in the search for exoplanets.

The early 2000s marked a golden age for finding exoplanets. Missions like NASA's Kepler Space Telescope, launched in 2009, transformed the field by providing a vast amount of data. Kepler observed over 150,000 stars, leading to the discovery of thousands of exoplanets, many of which were Earth-sized and located in their stars' habitable zones. These discoveries not only drove scientific exploration but also sparked public imagination as we began to ponder the possibilities of life beyond Earth.

With every new finding, the idea of discovering a twin to our beloved planet felt more achievable. The incredible variety of exoplanets found—from rocky worlds to gas giants, and from extreme climates to more temperate ones—ignited excitement. It felt as if we were engaged in a cosmic game of hide-and-seek, with the universe gradually revealing its secrets to us.

As the 2010s rolled on, the field continued to grow, fueled by technological advancements and global collaboration. New observatories and missions, both on the ground and in space, were launched specifically to expand our catalog of exoplanets. The Transiting Exoplanet Survey Satellite (TESS), launched in 2018, aimed to scan the entire sky in search of planets around the nearest and brightest stars. This ambitious endeavor built on Kepler's legacy and aimed to elevate the search for habitable worlds.

Throughout this journey, it became clear that the search for exoplanets went beyond academic interest. It was deeply connected to our quest to understand life itself. The big questions loomed large: What do we need for life to exist? Are there Earth-like planets where life could thrive? And if such planets exist, how might we ever communicate with any distant civilizations? Each discovered exoplanet added another piece to a larger puzzle, offering insights into our own existence and the possibility of life beyond our fragile home.

As astronomers began to study these far-off worlds more closely, the next logical step was to explore their potential for supporting life. Identifying Earth-like exoplanets became a primary goal, with scientists focusing on those that might have the right conditions for life as we know it. Key factors such as distance from their stars, atmospheric makeup, and surface conditions were examined in detail. The quest shifted from simply finding planets to discovering the right ones—those that could potentially offer a sanctuary for life.

The excitement surrounding exoplanet discoveries also led to increased funding and support for research in this field. Governments and private organizations began investing in missions aimed at searching for extraterrestrial life. A sense of urgency took hold, driven by the realization that each new discovery could be vital in answering our deepest questions about existence and life in the universe. The exoplanet revolution was not just a scientific pursuit; it became a cultural phenomenon, capturing the imaginations of people all over the globe.

Looking back on the evolution of exoplanet discovery, it's clear that this journey is just beginning. Each new finding raises even more questions and increases the stakes in our quest for knowledge. The mysteries that lie beyond our solar system still elude us, but with every technological advancement and every new exoplanet cataloged, we get a little closer to unraveling the secrets of the universe.

These discoveries hold significance that stretches well beyond scientific inquiry. They challenge us to rethink our place in the cosmos, reflect on the nature of life, and consider our responsibilities as caretakers of our planet. The exoplanet revolution encourages us to dream big and hope for more from the universe, to embrace the spirit of exploration, and to nurture a curiosity that reaches far beyond our own lives.

As we stand at the threshold of this new era in astronomy, there's a shared excitement among scientists and the public alike. The exoplanet revolution isn't just about uncovering new worlds; it's a journey of self-discovery, a quest to understand our origins and our ultimate destiny. The story of exoplanets is, in many ways, the story of humanity—a reflection of our natural drive to explore, question, and reach for the stars.

**Earth-like Exoplanets of Interest**

The search for understanding life beyond our blue planet has turned into an exciting journey, fueled by the hope of discovering Earth-like exoplanets scattered across the universe. So, what makes a planet "Earth-like"? While there isn't a single definition that everyone agrees on, a few key features usually take center stage: size, composition, distance from its host star, and the all-important presence of liquid water.

These factors are crucial in figuring out if a planet can potentially support life as we know it, and they guide our quest for cosmic neighbors.

Let's begin with size. The best candidates for Earth-like exoplanets are those that closely match our planet's physical dimensions. This similarity is important because if a planet is much larger or smaller than Earth, its atmosphere and surface conditions might be very different. A planet that's too small may struggle to hold onto an atmosphere, while a huge planet might resemble a gas giant, lacking the solid ground necessary for life.

Next up is the planet's composition. Ideally, an Earth-like exoplanet should be rocky, not gaseous. This distinction is crucial because rocky planets can support geological processes and have solid surfaces, which are essential for creating the right conditions for life. Additionally, the presence of key elements like carbon, oxygen, and nitrogen plays a significant role. These elements are the building blocks of life, providing the necessary foundation for complex organic molecules to form, which is where life starts to take shape.

Distance from the host star is also a vital aspect when identifying Earth-like planets. The habitable zone, often called the "Goldilocks Zone," is the area around a star where conditions might be just right for liquid water to exist. If a planet is too close to its star, it could get too hot and lose its atmosphere. Too far away, and it might be stuck in a deep freeze. Finding a planet in this sweet spot is like striking gold in the vast universe.

We can't forget about water, especially in its liquid form. It's often said that water is the essence of life. Our planet thrives thanks to its vast oceans, rivers, and lakes, which support countless ecosystems. So, confirming the presence of water—or at least the potential for it—on an exoplanet significantly increases its chances of hosting life. This brings us to some amazing candidates that have caught the eyes of scientists and dreamers alike.

Let's kick off our exploration with Proxima Centauri b, a planet that has sparked curiosity since it was discovered. Located just 4.24 light-years from Earth, Proxima Centauri b is the closest known exoplanet to our solar system. It orbits within the habitable zone of its red dwarf star, Proxima Centauri, and is about 1.17 times the size of Earth, making it a top contender for Earth-like status.

What makes Proxima Centauri b particularly fascinating is the type of star it orbits. Red dwarfs are known for their long lifespans but can also be unpredictable, often producing flares that could strip away the atmosphere of nearby planets. This raises an important question: Does Proxima Centauri b have an atmosphere that could support life? Recent studies suggest that, despite being bombarded by radiation, the planet may still have a protective atmosphere that could allow for liquid water to exist.

Additionally, the temperature on Proxima Centauri b seems just right for water to flow. However, much about its environment remains a mystery. The chances for life, whether it's simple microbes or something more complex, depend on many factors that scientists are still working to uncover. As our telescopes become more advanced and our technology improves, we may soon get a clearer view of what this neighboring world holds.

Continuing our journey through space, we come upon Kepler-186f, a planet that has made waves in the scientific community since its discovery in 2014. Roughly 500 light-years away, Kepler-186f is special for being the first Earth-sized planet found in the habitable zone of its star. This rocky world is about 1.1 times the size of Earth, making it a close match.

What really sets Kepler-186f apart is not just its size but where it sits in the habitable zone of a cooler, dim star known as a K-dwarf. This unique setup offers an exciting chance for scientists to consider how life might develop in different stellar environments. The star's lower brightness creates a more stable climate, protecting the planet from the intense radiation that would be likely around hotter stars.

The discovery of Kepler-186f has significant implications. It shows that Earth-like planets can thrive in various environments and that the potential for life isn't limited to sun-like stars. The excitement surrounding Kepler-186f lies in the possibility that it may have the right conditions for life, broadening our understanding of where we might find our cosmic relatives.

Our journey through the stars wouldn't be complete without exploring the TRAPPIST-1 system, a treasure chest of potential habitable worlds. This remarkable system, located about 40 light-years away, features seven Earth-sized planets, three of which sit comfortably in the habitable zone of their ultra-cool dwarf star. The sheer number of potentially habitable planets so close to each other is unprecedented and has captured the attention of astronomers around the world.

The standout candidates in the TRAPPIST-1 system are TRAPPIST-1e, f, and g, each presenting exciting possibilities for sustaining life. These planets are similar in size to Earth, boasting rocky compositions and perfect distances from their star, which might allow liquid water to exist on their surfaces. The potential for life on these planets is strengthened by the fact that they orbit a star much more stable than our sun, creating more temperate climates.

The TRAPPIST-1 system challenges our ideas about habitability, offering a unique situation where multiple planets might be able to support life within a single star system. This discovery fuels conversations among scientists about the nature of life and its adaptability. If life can thrive in such varied conditions, what does that suggest about the mysteries waiting for us in our galaxy? Are we on the verge of uncovering a whole new array of life forms?

As researchers continue to study these Earth-like exoplanets, they gather valuable insights that deepen our understanding of how planets form and what conditions are necessary for life. The discoveries of Proxima Centauri b, Kepler-186f, and the TRAPPIST-1 system not only add to our list of potential habitable worlds but also stretch the limits of our imagination.

Every exoplanet we investigate reminds us of the endless possibilities that lie beyond our solar system. With each new discovery, we get closer to answering fundamental questions about life in the universe. These Earth-like candidates strengthen the idea that we might not be alone, and the cosmos is bursting with opportunities yet to be explored.

Searching for Earth-like exoplanets is about more than just finding other worlds; it's a deep dive into our existence and the nature of life itself. The excitement surrounding these discoveries uplifts the human spirit, encouraging us to look beyond our own planet and consider the mysteries that await us in the universe. As technology continues to advance, so too will our understanding of these distant worlds—each star, each planet, a page in the grand story of existence, urging us to dream big and reach for the stars.

## Searching for Biosignatures

In the incredible stretch of the universe, where billions of stars shimmer like scattered jewels against a dark canvas, the search for life beyond our planet takes the spotlight. At the heart of this cosmic adventure is the idea of biosignatures—those intriguing clues that hint at the possibility of life, both familiar and yet to be discovered. But what are biosignatures, and how do we hope to recognize them from light-years away?

Biosignatures are essentially signals that indicate life has existed or might currently exist on another celestial body. These signals can be direct or indirect. Direct biosignatures might include specific organic molecules, while indirect ones often show up as changes or patterns in a planet's atmosphere that suggest biological processes are at play. For example, finding both methane and oxygen in a planet's atmosphere could point to life, as these gases usually react with each other and shouldn't be found together in large amounts without a way to replace them.

As we aim our telescopes at distant exoplanets, the methods we use to hunt for these biosignatures become vital. In this cosmic whodunit, scientists have a toolbox of techniques, with spectroscopy being one of the most powerful.

Spectroscopy involves analyzing the light spectrum that something—like an exoplanet—emits, absorbs, or scatters. When light from a star shines through a planet's atmosphere, it interacts with the gases present, creating a unique signature made up of different wavelengths. By studying this spectrum closely, scientists can identify specific gases that may hint at the presence of life.

Imagine holding a prism to sunlight and watching it split the light into a beautiful rainbow; this is essentially what spectroscopy does, just on a much grander scale. For exoplanets, the light we analyze comes from distant stars, illuminating the atmospheres of the planets that orbit them. Each gas has a distinct spectral fingerprint. For instance, if we find a lot of oxygen, it could be a major breakthrough since, on Earth, oxygen is mainly produced by photosynthetic organisms. The presence of methane, which is often produced by biological processes, could further support the idea of life.

However, it's not merely about spotting a single molecule; it's about understanding how different gases interact with one another and their various concentrations. Just finding oxygen alone doesn't guarantee life; we need to look at the bigger picture. This intricate dance makes the search for biosignatures a complex challenge that requires advanced technology and creative ideas.

Now, let's introduce the next key player in this story: space telescopes. These amazing tools act as our eyes to the universe, capturing light from distant worlds and bringing it into focus for examination. Among them, the James Webb Space Telescope (JWST) shines brightly as a source of hope and potential. With its launch on the horizon, the JWST is geared up to study the first galaxies that ever formed, but it also holds tremendous promise in our search for biosignatures.

Thanks to its powerful infrared capabilities, JWST can investigate the atmospheres of exoplanets, even when they're faint and far away. It can identify and analyze the chemical makeup of these atmospheres, looking for crucial gases that might indicate biological activity. Picture a futuristic detective armed with state-of-the-art technology, piecing together hints from the faintest whispers of light. That's the role space telescopes, especially JWST, play in our hunt for extraterrestrial life.

The JWST's ability to scan a wide range of wavelengths means it can uncover signatures that were once elusive. It will be able to examine planets located in the habitable zones of their stars—those sweet spots where conditions might be just right for liquid water, and potentially, life. The data it collects won't just enhance our understanding of specific exoplanets, but it will also broaden our knowledge of life's potential across the universe.

Still, with these exciting possibilities come significant challenges and ethical questions. The way we define life is a tricky puzzle that scientists and thinkers must carefully navigate. If we were to discover biosignatures, how would we interpret them?

Are we truly ready for the implications of finding life that doesn't fit our Earth-based understanding?

Moreover, searching for biosignatures prompts us to think deeply about what life really is. Life on Earth is based on carbon, relying on water and oxygen, but could there be other forms of life that operate on entirely different principles? Discovering biosignatures on distant planets could challenge our current definitions and assumptions, pushing the boundaries of what we believe life can be.

The ethical aspects of such discoveries are just as crucial. If we find clear signs of extraterrestrial life, what responsibilities do we hold not only for our planet but potentially for others as well? Should we try to communicate, or would that put both us and these distant beings at risk? The potential for life beyond Earth drives us to rethink our place in the cosmos and the consequences of our explorations.

Thinking about how humanity would react to finding biosignatures is a fascinating exercise. Many of us are familiar with science fiction stories where the discovery of aliens leads to either wild excitement or deep fear. Perhaps we would celebrate the fact that we are not alone, opening up new avenues in our understanding of biology and evolution. On the flip side, we might feel anxious—

wondering if they could be hostile or if they bring diseases.

Reflecting on history, every major scientific breakthrough has led to a reassessment of our role in the universe. The Copernican revolution, which revealed that Earth isn't the center of the cosmos, changed how humanity sees itself. Discovering biosignatures could launch us into a new era of understanding—one that recognizes us as part of a larger cosmic community. In this scenario, the implications extend beyond science; they reach into philosophy, religion, sociology, and ethics.

As researchers continue to improve their methods for detecting biosignatures, our imagination sparks with the potential of finding life beyond Earth. The techniques of spectroscopy and the advanced observations offered by space telescopes like the JWST are leading the charge in this quest. Each exoplanet we study isn't just a distant world; it becomes a window into understanding the possibilities of life in the universe and a chance to learn more about ourselves.

The search for biosignatures encourages us to look beyond our little blue planet, inviting us to contemplate the vastness of existence and our role within it. Every clue we uncover can lead to new questions and possibilities, and perhaps even new neighbors in this spectacular universe. As we stretch the limits of what we know, we are reminded of the human spirit's endless curiosity and desire to explore, connect, and ultimately understand. The stars call us forward, and as we venture deeper into the cosmos, we may discover that the true treasure lies not just in finding life, but in the questions that arise along the way.

# Chapter 6: The Search for Signals: The SETI Initiative

The question of whether we are alone in the universe has fascinated people for centuries, long before anyone even had a name for extraterrestrial life. This curiosity began to take on a more organized form in the 20th century with the creation of the Search for Extraterrestrial Intelligence, or SETI. As we take a closer look at SETI's journey and how it has changed over time, we'll explore the cultural roots, technological breakthroughs, and key individuals who helped transform this idea from mere speculation into a serious scientific pursuit.

To really understand where SETI came from, it's helpful to look at how humanity has interacted with the cosmos over the ages. Ancient civilizations often gazed up at the night sky, weaving stories about heavenly beings. The Greeks pondered the nature of the universe, while thinkers like Copernicus and Galileo laid the groundwork for a more scientific understanding of our place among the stars. Yet it wasn't until the 19th century that scientists began seriously considering the possibility of life beyond Earth.

In 1853, a British astronomer named William Whewell proposed the idea of the "plurality of worlds," suggesting that other planets might be inhabited. This idea caught on, sparking the imaginations of both scientists and dreamers. By the early 20th century, interest in finding extraterrestrial life was on the rise, with inventors like Nikola Tesla claiming they were receiving signals from Martians. Tesla's belief in the potential for communication with other worlds planted the seeds for what would eventually become SETI.

Jump ahead to the 1960s—a time marked by the Space Race and incredible technological progress. This was when the formal search for extraterrestrial intelligence truly began to take shape. In 1960, astronomer Frank Drake launched Project Ozma, named after the fictional princess from the Land of Oz. Using a small radio telescope, Drake listened for signals from two nearby stars, Tau Ceti and Epsilon Eridani. While the project didn't find any clear signs of life, it set the stage for future SETI efforts and introduced the now-famous Drake Equation. This formula aimed to estimate how many active, communicative extraterrestrial civilizations might exist in the Milky Way galaxy, giving both researchers and space enthusiasts something to grasp onto.

Another significant milestone came in 1984 with the founding of the SETI Institute, created by a group of scientists that included the well-known astrophysicist Jill Tarter. The institute aimed to deepen our understanding of life in the universe. Tarter became a leading figure in the SETI field, dedicating her career to searching for signals from other civilizations. Her passion and advocacy helped bring SETI into the spotlight, allowing scientists to openly pursue the possibility of life beyond Earth.

As time went on, SETI didn't just evolve in its methods; it also became more engaged with the public. In the early days, the scientific approach was somewhat closed off. However, in the late 1990s and early 2000s, initiatives like SETI@home allowed regular people to get involved. By downloading software that analyzed radio signals from space whenever their computers were idle, volunteers became a vital part of the search. This connection between science and the community sparked renewed interest in extraterrestrial life and opened up the conversation to a wider audience.

The technological advancements of the 21st century have given SETI a significant boost. Powerful telescopes like the Allen

Telescope Array allow scientists to scan large areas of the sky with remarkable accuracy. Additionally, the rise of big data analytics and machine learning has transformed how researchers sift through the overwhelming amount of data, searching for signals that could mean something. With new algorithms designed to identify patterns and noise, the search for extraterrestrial communications has entered an exciting new phase. Rather than waiting for a signal to appear, scientists can actively seek one out in ways that were unimaginable just a few years ago.

Despite these exciting developments, challenges still remain in the quest for signals from beyond. One of the biggest obstacles is sorting through the noise created by human-made radio emissions. With Earth buzzing with signals, finding a genuine extraterrestrial message becomes increasingly difficult. Scientists are continually refining their methods to ensure that any discoveries are credible and meaningful. As they tackle these challenges, the spirit of inquiry continues to shine, showcasing humanity's unwavering fascination with the universe.

Looking back at the history and evolution of SETI, it becomes clear that this initiative is more than just a scientific project; it's an exploration of existence itself. The search for extraterrestrial life invites us to think about our place in the universe and what life really means. Each step forward, each setback, and every story shared contribute to a larger narrative that connects us to the stars.

As we think about the future of SETI, it's crucial to realize that our pursuit of extraterrestrial life reflects our deepest dreams and wishes. The journey so far has been filled with hopes, challenges, and triumphs. The story of SETI is a testament to humanity's relentless quest to understand our own existence and to reach out into the unknown. As we keep our eyes to the skies, we are reminded that seeking life beyond Earth isn't just about finding signals; it's about building a connection that stretches across the cosmos.

In this ongoing effort, every scientist, amateur stargazer, and curious individual plays an important role in deepening our understanding of the universe. The history of SETI is about exploration, passion, and an unshakeable belief that we might not be alone in the vastness of space. And as we stare into the night sky, filled with countless stars and possible worlds, we can't help but wonder: what messages might be waiting for us in the silence of the cosmos?

## Techniques for Signal Detection

The vastness of space has always amazed and puzzled us. When we look up at the night sky, we can't help but wonder: are we really alone out there? Luckily, advancements in technology have opened up new paths for exploration, especially in the hunt for intelligent life beyond Earth. This adventure into the cosmos involves careful and thoughtful sorting through the flood of signals that come at us from all directions. To truly understand how researchers work in this field, it's helpful to learn about the basics of radio astronomy and the methods scientists use to detect possible signals from other forms of life.

Radio astronomy first emerged in the mid-20th century and takes advantage of the electromagnetic spectrum, which includes various wavelengths from radio waves to gamma rays. The beauty of radio astronomy lies in its ability to capture waves that travel freely through the vastness of space, often without being blocked by dust and gas that can hide visible light signals. These radio waves can come from natural celestial events—like pulsars, quasars, and the cosmic microwave background radiation—or from engineered sources, such as signals that might be sent out by advanced civilizations.

At the center of radio astronomy are radio telescopes. These work similarly to optical telescopes but are designed to catch radio waves instead of visible light. Imagine a giant dish, much like a large satellite dish, that collects incoming radio waves and channels them into sensitive receivers. These receivers then convert the waves into electrical signals that scientists can analyze for patterns and details. Technology has come a long way since the early days, and today's radio telescopes are equipped with sophisticated arrays and advanced software that boost their sensitivity and accuracy.

One of the most impressive setups in radio astronomy is the use of large radio arrays. For example, the Very Large Array (VLA) in New Mexico consists of 27 antennas arranged in a Y-shaped pattern. Each antenna works together like a giant single telescope, allowing researchers to see a large area of the sky with high clarity. This teamwork makes it easier for scientists to pick up on faint signals, making it a key tool in the search for extraterrestrial messages. Similarly, the Green Bank Telescope in West Virginia, one of the largest movable radio telescopes globally, offers remarkable sensitivity to weak radio emissions, enhancing scientists' chances of detecting signals from distant civilizations.

While radio waves are a major focus for the Search for Extraterrestrial Intelligence (SETI), researchers also look into optical signals. Optical SETI uses telescopes to search for laser pulses or other types of light that could signal intelligent communication. Just like radio waves can travel across the universe, photons from laser transmissions can cover vast distances if directed well. Using light for communication isn't just a fantasy; it's similar to the technological progress we see on Earth, where fiber optic networks have changed the way we transmit data. Therefore, optical SETI provides another valuable path in the quest to find extraterrestrial intelligence.

To fully appreciate the complexities of signal detection, it's vital to understand the electromagnetic spectrum. Different wavelengths of electromagnetic radiation provide distinct information about celestial objects. For example, radio waves have longer wavelengths than visible light and can better penetrate clouds of interstellar dust. On the flip side, optical signals, which have shorter wavelengths, can give us insights into the chemical makeup of faraway stars and planets. The choice of wavelength is crucial, as it significantly impacts the type of information researchers can gather from their data.

Within this detailed world of radio and optical astronomy lies the idea of the signal-to-noise ratio (SNR). Finding extraterrestrial signals can feel like trying to hear a whisper in a busy crowd. The SNR measures how strong a desired signal is compared to the background noise, with a higher ratio meaning a clearer signal. To differentiate potential extraterrestrial signals from natural sources and human-made noise, researchers use advanced filtering techniques. Without these methods, the overwhelming noise of radio emissions from satellites, cell towers, and other sources would make the search nearly impossible.

The challenges of signal detection go beyond just the technical abilities of the instruments. Scientists must also deal with the massive amounts of data produced by radio telescopes and the constant flow of noise. As these telescopes scan the universe, they generate terabytes of information. Sorting through this massive amount of data to find a real signal is like looking for a needle in a haystack. To help with this, researchers have turned to machine learning and advanced algorithms, which have changed how signals are analyzed. These technologies can spot patterns in the data that might suggest the presence of a signal, making it easier to identify potential communications.

Throughout the history of SETI, many studies and projects have aimed to detect signals from beyond our planet. One of the most famous examples happened in 1977 when astronomer Jerry R. Ehman, working at Ohio State University, picked up a strong radio signal that lasted for 72 seconds. The signal was so remarkable that he circled it on the data printout, writing "Wow!" in the margins. This moment, known as the Wow! signal, has become iconic in the search for extraterrestrial intelligence. Despite thorough investigations, the source of the signal has never been definitively identified, leaving it as an intriguing mystery in the ongoing quest for cosmic communication.

The Wow! signal captures the spirit of what drives scientists in the field of SETI. The dedication, curiosity, and relentless pursuit of knowledge are essential traits for those navigating the complexities of signal detection. The search for signals from intelligent civilizations goes beyond just a scientific quest; it's an exploration of the unknown, reflecting the hopes and dreams of countless individuals who long to connect with something larger than themselves. For every success, there are stories of setbacks and disappointments, yet the drive to explore remains strong, pushing researchers to keep searching.

On top of the technical challenges, there are important ethical questions that arise in the search for extraterrestrial signals. As scientists and enthusiasts think about what it would mean to discover intelligent life beyond Earth, they wonder how humanity would respond to such a revelation. Would we be ready to interact with a civilization that might have technologies far beyond our own? The potential for unexpected consequences is significant, reminding us that seeking knowledge often comes with responsibilities we must think through carefully.

As researchers keep improving their techniques for signal detection, the field of SETI continues to grow and change. By combining cutting-edge technologies with traditional methods, we are entering a new age of exploration. The rise of high-speed computing, along with new observation techniques, has shifted the search for extraterrestrial signals from a passive activity to an active and engaging pursuit. The conversations around the implications of these advancements highlight the need to balance scientific inquiry with the ethical questions that arise as we expand our understanding of the universe.

In the grand picture of scientific exploration, the search for extraterrestrial signals shows humanity's endless curiosity and ambition. Each study, every signal analyzed, and all the surprising discoveries contribute to a larger story that connects us to the cosmos. The question of whether we are alone is more than just a scientific query; it's a timeless journey that ties us to the mysteries of existence. As we keep looking up at the stars, we open ourselves to the possibilities of the universe, hoping that one day a message from the cosmos might reach us. The pursuit of extraterrestrial intelligence isn't just about finding answers; it's about embracing the wonder and mystery that the universe offers, reminding us that our desire to connect is as profound as the cosmos itself.

## The Future of SETI: New Technologies

As we step into an exciting new chapter in the Search for Extraterrestrial Intelligence (SETI), it's thrilling to see how fresh technologies are transforming our mission to uncover the mysteries of the universe. The tools we have today are advancing like never before, thanks to innovations in artificial intelligence, machine learning, and cutting-edge observational instruments. These developments are opening doors to discoveries that once felt like the stuff of science fiction.

Artificial intelligence has moved beyond being just a fancy idea for the future; it's now a key player in how we sift through the enormous amounts of data that radio telescopes and other observatories produce. The sheer scale of information collected in the hunt for signals from other worlds can be staggering. As the number of observations skyrockets, it becomes trickier for human researchers to filter out the noise and pinpoint possible signals. That's where AI and machine learning come into play, allowing us to analyze massive datasets at incredible speeds.

Think about searching for a favorite song in a library filled with millions of records. A person could easily spend days or even weeks looking for it, while an AI algorithm could do the same job in just seconds. These smart algorithms are designed to spot patterns and oddities that might escape the human eye. They can learn to tell the difference between random cosmic events and signals that hint at intelligent life. This boosts our chances of picking up extraterrestrial signals since researchers can focus on the most promising leads instead of being overwhelmed by the background noise of the universe.

But it's not just about speed; it's also about getting things right. With machine learning, the more data the models process, the better they get. This means researchers can sharpen their search criteria and have a clearer picture of what potential signals from extraterrestrial civilizations might look like. We're no longer limited to hunting for specific frequencies or patterns; instead, we can explore a wider range of possibilities, paving the way for new discoveries.

One of the most ambitious initiatives on the horizon is the Square Kilometre Array (SKA), a groundbreaking radio telescope project that could revolutionize astronomy and SETI. Once it's fully up and running, the SKA will be the largest radio telescope on the planet, spanning an area of one square kilometer. This massive structure will be set up across various sites in Australia and South Africa, harnessing the combined power of thousands of antennas. The SKA will significantly boost our ability to detect extraterrestrial signals, allowing us to pick up much fainter signals than ever before.

The SKA represents an incredible global teamwork effort, with contributions from countries all around the world. This collaboration highlights how essential it is to work together in our quest for cosmic knowledge. Scientists, engineers, and curious minds from diverse backgrounds are joining forces, sharing their skills and resources to expand our understanding of the universe. The SKA is not just a technological achievement; it embodies humanity's deep desire to understand our place in the cosmos and connect with possible extraterrestrial beings.

Moreover, the advancements from the SKA will let researchers conduct more thorough surveys of the universe. They won't just be looking for signals from extraterrestrial intelligence; they'll also gather vital information about how galaxies form and evolve, the behavior of black holes, and the nature of dark matter. This treasure trove of data will deepen our understanding of the cosmos and help us put any potential signals we might receive into context.

As we gaze toward the stars, we must also recognize the role of space-based observatories and missions in the future of SETI. The recently launched James Webb Space Telescope (JWST) is set to change the way we see the universe. Located beyond Earth's atmosphere, the JWST can observe cosmic phenomena without interference from atmospheric noise. This clarity will not only allow us to study distant stars and planets more effectively but will also increase our chances of spotting signs of life.

The benefits of these space-based observatories go beyond just better observations. By studying exoplanets—those planets that orbit stars outside our solar system—scientists can look for biosignatures, which are clues that life might exist. The JWST will enable researchers to analyze these distant worlds' atmospheres, searching for chemical markers that could hint at life. While finding signals from intelligent civilizations is the ultimate goal of SETI, the search for microbial life is equally fascinating and crucial for understanding the universe.

However, discussing new technologies in SETI also brings up important philosophical questions. As we create more advanced ways to search for extraterrestrial life, we must think deeply about what communication really means. What if we encounter lifeforms that are entirely different from us—beings that don't communicate through sounds or light but rather through methods we haven't even imagined yet?

This idea challenges our traditional definitions of intelligence and civilization. For ages, we've defined intelligence based on human traits: language, technology, and culture. But as we broaden our exploration of the cosmos, we may find forms of life that challenge these definitions. They might communicate through vibrations, chemical signals, or even electromagnetic patterns that are currently beyond our ability to perceive. The act of searching for extraterrestrial intelligence pushes us to confront our biases and expand our understanding of what it means to be "intelligent."

As we push the limits of what we know and can do, we must also consider the ethical implications of what we might discover. If we were to receive a signal from an extraterrestrial civilization, how should we respond? What responsibilities would we have as caretakers of our planet and representatives of humanity? These questions require thoughtful discussion as we navigate the complexities of our cosmic journey.

Despite these challenges, the future of SETI is undoubtedly hopeful. With each technological breakthrough, every international collaboration, and each new finding, we come closer to the possibility of connecting with intelligent life beyond our planet. Humanity's unquenchable curiosity and relentless quest for knowledge drive this exciting endeavor. The spirit of inquiry—paired with a deep-seated belief in the unknown—fuels our desire to explore the cosmos and reach out to life beyond Earth.

In this remarkable era of technological advancement, we are equipped with tools that allow us to explore the very foundations of our universe, uncovering secrets that have been elusive for generations. The possibility of discovering extraterrestrial intelligence isn't just a scientific goal; it embodies the heart of human exploration. It captures our dreams, aspirations, and our longing to connect with something greater than ourselves.

As we look forward to the future of SETI, we are reminded of the wondrous mysteries waiting for us in the cosmos. The tools we have—whether they are advanced algorithms, massive radio arrays, or space-based observatories—empower us to explore realms of existence that were once just dreams. Every signal we analyze, every data point we collect, brings us closer to the answers we seek. In this grand adventure, we celebrate our humanity, our curiosity, and our relentless drive to understand the universe and our role in it.

Ultimately, the search for extraterrestrial intelligence is not just about uncovering answers; it's about embracing the wonder and possibilities that the universe holds. It's a celebration of our shared existence and a collective journey into the unknown, reminding us that the cosmos is vast, and we are just beginning to scratch the surface of its enigmas. As we continue to look to the stars, we do so with open hearts and minds, eager to welcome whatever wonders may lie beyond our home planet. The quest for life beyond Earth is a journey that transcends borders, cultures, and time—a journey that invites everyone to join in exploring the cosmos and seeking connections with the unknown.

# Chapter 7: Life's Extremes: Lessons from Earth

When we think about life on Earth, we often picture lush forests, rolling hills, and vibrant coral reefs. These beautiful places, full of different kinds of plants and animals, seem to show us the best of what life can be. But if we look a little deeper, we'll find an amazing world that challenges everything we know about life. This world is home to some truly incredible organisms called extremophiles. These tiny survivors thrive in environments that would scare most other living things. They remind us just how strong and adaptable life can be, even when faced with the toughest challenges.

Extremophiles live in places that are extreme in all sorts of ways. You can find them in boiling hot springs, where temperatures climb way above what we would even dare to handle. They also thrive in acidic lakes that could eat through normal biological structures and in the freezing depths of the ocean, where temperatures drop well below freezing. These organisms have not just managed to survive in these harsh conditions; they've actually learned how to thrive. The term "extremophile" tells us a lot: "extremo" means extreme, and "phile" means lover of. So, these are creatures that not only endure extreme situations but actually prefer them!

Studying extremophiles is fascinating not just for scientists but also for anyone curious about life beyond our planet. If these strong organisms can live in such unwelcoming places on Earth, what might that mean for life on other planets or moons that might have their own challenging environments? As we explore space, extremophiles can help us imagine what kind of life forms we might find on celestial bodies that have their own unique challenges.

One of the most famous groups of extremophiles is thermophiles. These are heat-loving organisms found in places like geysers and hot springs, where temperatures can soar to a scorching 80 degrees Celsius (176 degrees Fahrenheit) or even higher. At first glance, it seems impossible for life to exist in such intense heat. But thermophiles have adapted in incredible ways. They have special proteins and cell structures that allow them to thrive in these high temperatures. Their enzymes, known as thermostable enzymes, work best when it's hot, making them valuable in various industries, including biofuels and pharmaceuticals.

On the opposite end of the temperature spectrum, we have psychrophiles, the cold-loving cousins of thermophiles. These hardy organisms make their homes in icy places like glaciers, polar ice caps, and the chilly waters of the deep sea. Psychrophiles have special adaptations that let them survive in temperatures that are frequently at or below freezing. They manage to keep their cell membranes flexible, which would normally become stiff in such cold. One amazing adaptation they have is antifreeze proteins, which stop ice crystals from forming inside their cells, allowing them to survive in the frozen world around them.

Then there are halophiles, the salt-loving organisms that thrive in super salty environments like salt flats, salt mines, and even the Dead Sea, where salt levels are so high that most living things would dry out and perish. These incredible creatures have developed ways to handle the extreme pressure that comes from all that salt. One fascinating thing about halophiles is their ability to gather a chemical called glycerol in their cells, helping them balance the pressure from the salt and preventing dehydration. They don't just survive in these salty places; they embrace them, turning what looks like a barren wasteland into a home.

As we explore the world of extremophiles further, it's clear that their unique traits help them thrive where life seems impossible. But this raises an interesting question: if life can exist in these extreme places on Earth, what other kinds of life might be out there in similar extreme conditions on other planets? Studying extremophiles opens up a world of possibilities and encourages us to think about extraterrestrial life.

Take the icy moons of Jupiter and Saturn, like Europa and Enceladus, for example. Underneath their frozen surfaces is thought to be a warm ocean of liquid water, kept that way by internal heat. While the conditions may be harsh, the presence of water—a vital ingredient for life—offers exciting possibilities. Extremophiles could give us clues about what life might look like in these alien oceans. Just imagine: tiny, resilient organisms similar to those on Earth could be swimming in the waters of a distant moon, far from the warmth of our sun.

The potential doesn't stop there. Think about Mars, the red planet. It might look barren and cold today, but scientists have found ancient riverbeds and signs of past water, suggesting it used to be wetter and could have hosted life. Extremophiles give us hope that if life ever existed on Mars, it could have taken forms that challenge our traditional ideas of what life should be. Exploring Mars is not just about uncovering its history; it's also about discovering new forms of life.

In a universe full of possibilities, extremophiles show us that life can be incredibly resilient. They teach us that life isn't just about thriving in nice conditions; it's about adapting, bouncing back, and finding a way to exist against all odds. This perspective invites us to rethink what we believe life can look like and how it might exist. It encourages us to keep an open mind as we search for signs of life beyond our planet.

The story of extremophiles is a celebration of life's amazing ability to adapt and thrive in unexpected places. These organisms are more than just survivors; they are explorers of extreme environments that many people think are unlivable. They challenge our understanding of life and what it can become, both on Earth and out in the cosmos. As we look deeper into the universe for signs of life, it's these extraordinary extremophiles that might help us unlock some of its biggest mysteries.

By celebrating these amazing creatures, we also honor the spirit of discovery that drives us to learn more about our world. Each new discovery about extremophiles pushes us forward, inspiring future generations of scientists, explorers, and dreamers to think about what lies beyond the stars. The universe is vast and mysterious, and with every new finding, we get a little closer to understanding where we fit in. This exciting journey not only encourages us to explore our own world but also to dream big—because if life can thrive in the extreme conditions we see on Earth, just imagine what wonders await us in the cosmos!

## Extreme Habitats on Earth

Imagine diving into the dark ocean, where sunlight slowly disappears and pressure surrounds you like a heavy blanket. In this deep blue world, life thrives in places most folks would consider too harsh to support any living creatures. Take deep-sea hydrothermal vents, for example: these hot springs on the ocean floor shoot out water so hot it can reach temperatures of up to 400 degrees Celsius (752 degrees Fahrenheit). The water is rich in minerals, creating a dramatic contrast to the icy surroundings. Surprisingly, these extreme conditions are home to a wild variety of creatures, from enormous tube worms to colorful communities of bacteria. Once thought to be lifeless, these habitats now show us how incredibly resilient life can be, with organisms known as extremophiles not just surviving but thriving where others would fail.

Deep-sea hydrothermal vents act like underwater geysers, pouring out scalding water and supporting life against all odds. Here, tube worms and other extremophiles have found a way to harness energy from the raw materials around them through a process called chemosynthesis. Tube worms have a special relationship with bacteria that live inside them, which help convert the vent's minerals into energy. In this environment, devoid of sunlight—which is essential for most life on Earth—these tube worms challenge our understanding of how life can exist.

    The beauty of these ecosystems is breathtaking. Bright colors like deep reds, vivid yellows, and dark blues flash across the landscape, thanks to the diverse creatures that call this extreme habitat home. Tube worms can grow up to three meters long (nearly ten feet), standing tall with their striking red plumes swaying gently in the water. These plumes are not just for show; they serve an important purpose. They contain hemoglobin, which allows the worms to capture hydrogen sulfide, a toxic yet vital compound for their survival. It's hard to believe that life can be so vibrant and complex in such a forbidding place.

    Sharing this underwater garden with the tube worms are other extremophiles, like methanogens and various types of archaea.

These tiny organisms have mastered the ability to flourish in high-pressure, high-temperature environments where oxygen is scarce. For instance, methanogens turn carbon dioxide and hydrogen into methane, a process that not only helps them survive but also plays a role in the global methane cycle. The relationship between these organisms and their harsh surroundings shows us how life can adapt and find balance, even in places that seem hopeless.

Now, let's shift our focus to the warm embrace of acidic hot springs. Yellowstone National Park is a perfect example, where vibrant colors swirl in microbial mats, creating a stunning display that looks like nature's own artwork. The hot springs here can reach temperatures over 80 degrees Celsius (176 degrees Fahrenheit) and have acidity levels that rival battery acid. Shockingly, this harsh environment is bustling with extremophiles that have adapted to thrive in boiling, acidic waters.

The striking colors you see in these hot springs come from different types of bacteria, each perfectly suited to specific temperature and pH levels. Take Thermus aquaticus, for instance; this thermophilic bacterium is famous not only for its toughness but also for producing a special enzyme that's widely used in the biotechnology world. This enzyme, known as Taq polymerase, is vital for a technique called polymerase chain reaction (PCR), which has changed the field of molecular biology.

The microbial mats in these hot springs serve as a reminder that life can flourish in conditions that would make most organisms shudder. The high temperatures and acidity are not just obstacles to overcome; they are the specific conditions that shape these extremophiles into the resilient beings they are. In this lively ecosystem, extremophiles are not merely holding on; they are thriving, showcasing how adaptable life can be.

Next, let's travel to the chilly landscapes of the polar ice caps and glaciers. In the stark, frozen expanses of Antarctica and the Arctic, life stands as a remarkable example of endurance. The extreme cold and constant darkness create challenges that would be impossible for most living things. Yet, in this unforgiving environment, psychrophilic organisms, or cold-loving extremophiles, have found a way to exist, showing us that they can not only survive but also thrive.

Psychrophiles are specially adapted to freezing temperatures, often hovering just above the freezing point of water. They have unique features, like antifreeze proteins that prevent ice crystals from forming inside their cells, which would cause damage. Picture tiny organisms swimming in a sea of ice, their resilience shining through in the harsh conditions. Some psychrophiles can even metabolize at temperatures as low as -12 degrees Celsius (10 degrees Fahrenheit), highlighting the incredible adaptability of life.

The polar regions are more than just barren landscapes; they are vibrant ecosystems brimming with life. Beneath the ice, a variety of extremophiles, from bacteria to algae, contribute to the ecosystem's delicate balance. These organisms play a crucial role in carbon cycling and nutrient availability, proving that even the harshest environments can support complex interactions among living things.

Now, let's turn our attention to the dry expanses of saline lakes and salt flats. Places like the Great Salt Lake and the Dead Sea showcase extreme salinity that poses challenges even for the toughest organisms. However, halophiles, or salt-loving extremophiles, have evolved to thrive in these hyper-saline environments, transforming what many view as inhospitable into a lively haven of life.

In these saline lakes, halophiles have developed specialized cell structures to handle the intense pressure from the surrounding salt.

They produce substances like glycerol to prevent dehydration, allowing them to maintain their cellular health. These adaptations are not just fascinating; they highlight the clever strategies life employs to overcome extreme conditions.

The colors of these saline environments can be mesmerizing, with vibrant pinks and reds appearing thanks to carotenoid pigments in certain halophiles. While these tiny organisms are often invisible to the naked eye, they contribute to the unique beauty of these landscapes, reminding us of the hidden wonders found in extreme habitats.

Whether we're exploring the depths of the ocean, the fiery landscapes of hot springs, the icy tundras of the poles, or the desolate salt flats, it's clear that extremophiles are champions of adaptability. They show us that life isn't limited to comfortable settings; it can thrive in the most surprising places. Each of these extreme habitats tells a story of resilience and innovation, emphasizing the incredible diversity of life on Earth.

Moreover, studying these extreme environments has significant implications beyond our own planet. As we search for signs of life in the universe, understanding how extremophiles thrive in harsh conditions can offer valuable insights into the possibility of life elsewhere. If organisms can survive and thrive in Earth's extreme environments, what might we find on distant planets or moons?

Consider Europa, one of Jupiter's moons, with its icy surface hiding a vast ocean of liquid water kept warm by geothermal activity. This environment, while extreme, shares similarities with Earth's deep-sea hydrothermal vents, suggesting the potential for life forms adapted to these conditions. As we aim our telescopes and spacecraft toward the stars, the lessons we learn from extremophiles might help guide our quest for extraterrestrial life, shining a light on what possibilities await us.

In a universe filled with mysteries, extremophiles remind us that life is more adaptable than we often think. These remarkable organisms embody resilience, encouraging us to rethink what constitutes a viable habitat. They challenge us to look beyond the familiar and embrace the extraordinary variety of life that exists in the most unexpected corners of our planet.

Reflecting on the lessons these incredible organisms teach us reminds us of the beauty and complexity of life on Earth. Each extreme habitat is filled with stories waiting to be discovered, and each extremophile stands as a testament to the creativity of evolution. Exploring these environments continues to inspire scientists and dreamers alike, sparking curiosity about the intricacies of life and the endless possibilities that lie beyond our world.

In this grand story of extremophiles, we celebrate life's tenacity and are invited to embrace the unknown. The journey through Earth's extreme habitats not only uncovers the richness of our planet but also prompts us to ponder the larger mysteries of the universe. With every new discovery, we come closer to understanding our place in this vast cosmos, driven by the same curiosity and desire to explore that have always characterized humanity.

**Implications for Extraterrestrial Life**

The idea of life is a fascinating mystery that has intrigued people for ages. When we look up at the stars, it's natural to wonder if we're the only ones here in this enormous universe. The study of extremophiles—amazing organisms that manage to live in some of the hardest places on Earth—helps us expand our idea of what life can be. These unique creatures encourage us to think about "life as we don't know it," hinting that the forms of life we might find beyond our planet could be beyond anything we've imagined so far.

Astrobiology, the science that investigates the possibility of life beyond Earth, has gained a lot of insights from studying extremophiles. These tough little organisms show us how adaptable life can be, even when faced with the harshest conditions. Just like they've found ways to survive in boiling hot springs, frozen landscapes, and the deepest parts of the ocean, we can consider that life could also thrive in extreme environments on other celestial bodies. The search for extraterrestrial life is not just a matter of curiosity; it's a fundamental quest that speaks to our very existence, reminding us that life can take on many unexpected forms.

Take Mars, for instance, the most well-known candidate in our search for alien life. This reddish planet has intrigued scientists and dreamers for a long time. Evidence from various missions hints at the presence of underground water, raising exciting possibilities for the existence of life there. Just as extremophiles on Earth thrive in salty or acidic surroundings, Martian microbes, if they are out there, might have similarly adapted to live beneath the planet's barren surface.

Now think about Europa, one of Jupiter's most fascinating moons. Underneath its thick icy shell lies a vast ocean that may provide the right conditions for life. If we look at our own planet, we can see deep-sea hydrothermal vents teeming with life that thrives in complete darkness, feeding on the minerals that bubble up from the ocean floor. Europa's hidden ocean could replicate this extreme environment, prompting us to wonder what kinds of extremophiles might live beneath the ice. What special traits would they need to survive in a frozen world? Would they have anything in common with the tube worms and methanogens that thrive in our deep seas, or would they be completely different?

As we explore exoplanets—those distant worlds that orbit stars outside our solar system—we should keep our minds open about the kinds of life that might exist there. Some of these planets sit in what's known as the "Goldilocks Zone," where conditions are just right for liquid water to exist. Others, however, might be exposed to extreme temperatures, intense radiation, or crushing pressure. Just as halophiles have found ways to thrive in highly salty environments, life on an exoplanet could have developed unique adaptations to survive in tough conditions. The variety of extremophiles we've found on Earth serves as a crucial reference point in our search for signs of life on distant worlds.

What we learn from extremophiles goes beyond just showing that life can exist in extreme conditions. They challenge our traditional definitions of life and push us to rethink what is necessary for survival. In our quest to find extraterrestrial life, we must recognize that our definitions might not capture the many possibilities waiting for us in the universe. What if the life we are searching for isn't carbon-based or doesn't depend on water in the way we know? Extremophiles remind us that life can be resilient, adaptable, and ingeniously resourceful, thriving in environments we would consider hostile.

These questions concern more than just scientists; they have deep implications for all of humanity. As we think about the possibility of extraterrestrial life, we also reflect on our own place in the cosmos. The idea that life can thrive in the most unexpected places inspires a sense of wonder and humility. It encourages us to explore the universe with curiosity, seeking not just other beings but also understanding what it means to be a part of this grand story of existence.

When we consider what extremophiles mean for astrobiology and the hunt for extraterrestrial life, we should ask ourselves: What if we're on the verge of discovering that life is not only tough but also incredibly diverse? What if the life forms that exist elsewhere are so different from us that we can't even recognize them with our current ideas? The possibilities are both thrilling and daunting, pushing us to dream bigger and think more widely.

The adaptability of extremophiles highlights how life is shaped by its environment. The harsh conditions in extreme habitats on Earth have molded the organisms that can survive there. This connection between life and its surroundings raises questions about how extraterrestrial life might adapt. Would Martian organisms develop unique biochemical pathways to save water? Would life on Europa create antifreeze-like proteins to endure the chilling ocean beneath its icy crust? The answers to these questions are both exciting and mysterious.

These ideas matter especially when we think about the future of humanity. If we aim to become a space-faring civilization, we need to recognize that our survival may rely on our ability to adjust to new worlds, just like extremophiles have adapted to their tough environments. Exploring other planets and moons not only deepens our understanding of the universe but also makes us think about the resilience of our own species. We are not just seekers of knowledge; we are also caretakers of life itself, responsible for protecting and nurturing the delicate balance that sustains us.

As we journey deeper into the cosmos, the curiosity awakened by extremophiles will guide our exploration. Each discovery brings us closer to solving the mysteries of life beyond Earth. The search for extraterrestrial life isn't just a scientific pursuit; it also connects to humanity's deepest questions about existence and our role in the universe.

Studying extremophiles pushes us to think about life in all its forms, encouraging us to stay open-minded as we explore the far reaches of our solar system and beyond. It invites us to imagine a universe where life showcases a stunning variety of adaptations, a cosmic celebration of resilience and mystery. The story of extremophiles isn't just about survival; it's about the beautiful interplay between life and the universe.

So, as we gaze at the stars, let's draw inspiration from the tenacity and creativity of life on Earth. Let's embrace the unknown, driven by curiosity and wonder about the opportunities that await us in the universe. Every question we ask stirs our desire to explore, discover, and connect with the broader story of existence. The journey into the unknown isn't only about finding new worlds; it's also about understanding what it means to be alive in this vast universe.

Ultimately, the search for extraterrestrial life is a quest that resonates deeply with our humanity. It urges us to look beyond what we know and welcome the extraordinary. As we uncover the mysteries of life on Earth, we're reminded that we are just one thread in a complex and beautiful web of existence. The universe is immense, and our journey has only just begun. With every new discovery, we move closer to understanding not just the nature of life beyond our planet, but also the profound connections that bind us all together.

Miles Kepler

# Chapter 8: Sci-Fi and the Search for Extraterrestrial Life

The glowing screen flickers, casting a star-speckled sky that feels alive with possibilities. For many, this celestial display is more than just a pretty picture; it's a doorway to amazing worlds beyond our own. Since the dawn of time, people have looked up at the stars, wondering about the mysteries of the universe. Yet, it's through science fiction that our imaginations truly take off. This genre acts as both a mirror reflecting our own lives and a lens through which we explore our hopes, fears, and dreams about life beyond Earth. It holds the power to shape how we think and even influence real scientific efforts.

From the moment H.G. Wells introduced readers to Martians in "The War of the Worlds," the idea of life beyond our planet has enchanted people, sparking curiosity and debate among scientists and thinkers alike. Wells's thrilling tale blended adventure with deep social insights, prompting readers to ponder some big questions about what it means to be human and where we fit in the universe. Would we seek to conquer or be conquered? Would we open our arms in welcome or raise our fists in defense? Such questions echo throughout the world of science fiction, stirring a collective curiosity that goes well beyond the pages of books.

As we explore the vast universe of science fiction in literature and film, it's clear that this genre has been a rich source of ideas about extraterrestrial life. Arthur C. Clarke's "2001: A Space Odyssey," both a novel and a remarkable film directed by Stanley Kubrick, shows us a vision of humanity's growth tied to mysterious alien intelligence. The monoliths in the story stand for a turning point in human evolution, suggesting that our story—a brief flicker in the grand scheme of the universe—has been shaped by forces we can barely comprehend. This idea, that we might be standing on the shoulders of cosmic giants, makes us think about our own place in the larger story of existence.

Science fiction isn't afraid to tackle the big questions that come with meeting extraterrestrial beings. Philip K. Dick's thought-provoking works, like "Do Androids Dream of Electric Sheep?"—which inspired the movie "Blade Runner"—force us to think about what consciousness really is and what it means to be "alive." Are creations of humans—like androids, robots, or genetically modified beings—worthy of the same respect as organic life? Through these stories, science fiction encourages conversations that bridge the gap between imagination and science.

The influence of science fiction goes beyond books and spills into film and TV, too. The iconic series "Star Trek" presents an uplifting view of a diverse universe and has become a landmark of culture. Captain Kirk, Spock, and their crew venture where no one has gone before, exploring new worlds and meeting alien civilizations. The series introduced ideas like the Prime Directive, which focuses on not interfering in other species' lives. This storytelling framework invites viewers to think about the ethical questions surrounding contact with alien life. Because of this, "Star Trek" has inspired countless scientists and engineers, many of whom have gone on to careers in aerospace and astrobiology.

The impact of science fiction can also be seen in the increasing public interest in the search for extraterrestrial life. As novels and films showcase incredible beings from far-off galaxies, they spark real scientific curiosity about whether we are truly alone in the universe. The SETI (Search for Extraterrestrial Intelligence) initiative, which aims to pick up signals from intelligent alien civilizations, has gained support from both scientists and everyday people. The thrill of the unknown, combined with the influence of science fiction, drives the desire to explore the cosmos and seek answers to that age-old question: "Are we alone?"

As technology moves forward and our understanding of the universe grows deeper, the themes in science fiction become more important than ever. The discovery of exoplanets—planets orbiting stars outside our solar system—has opened new doors for astrobiologists. The thought that these distant worlds might host life has shifted from mere fantasy to an exciting possibility grounded in scientific exploration. Books like Kim Stanley Robinson's "Red Mars" series not only entertain us but also speculate on the future of human colonization of other planets, stirring conversations about sustainability, ethics, and how we adapt. This blend of imagination and real-world science shows how science fiction can inspire actual scientific progress.

Interestingly, the way we depict extraterrestrial life in science fiction has changed over the years. Early portrayals often mirrored societal fears—Martians were painted as evil invaders, while aliens in other stories were often seen as terrifying or grotesque. However, as society has evolved, so have our portrayals of alien beings. More recent works, such as "Arrival," based on Ted Chiang's short story "Story of Your Life," offer a more complex view. The film explores communication and understanding, suggesting that encounters with aliens don't have to be hostile. Instead, it presents the idea that these interactions could be about empathy, connection, and the exchange of knowledge.

The ability of science fiction to shape our views on extraterrestrial life is truly powerful. It inspires us and serves as a cautionary tale, reminding us of our actions' consequences, the fragility of existence, and the need for understanding. In a world where science is evolving rapidly, the stories we consume mold our attitudes and shape our dreams. The tales we tell about alien life resonate with our deepest hopes and fears, pushing us to think about the bigger picture of our curiosity.

As humanity seeks to uncover the universe's secrets, we must recognize how science fiction serves as a valuable tool that goes beyond simple entertainment. It sparks curiosity, broadens our perspectives, and encourages us to ask crucial questions about our existence. In many ways, we are all explorers on this cosmic adventure, diving into the uncharted territories of our thoughts and the universe itself. Science fiction invites us to dream, to be curious, and to reach for the stars, nurturing a connection to the cosmos that is fundamental to who we are.

Ultimately, as we think about the possibility of life beyond our planet, we can't help but look not only to the stars but also into the depths of our own imaginations. The stories spun through science fiction have enriched our understanding of the universe, allowing us to dream of the unimaginable and challenge us to journey into the unknown. Through these compelling narratives, our hopes and aspirations blend with the cosmos, shaping how we perceive what it means to be human in an infinite universe. This journey is filled with wonder, curiosity, and the never-ending quest for understanding—one that promises extraordinary discoveries yet to come.

## From Fiction to Science: Inspirations and Innovations

The fascinating connection between science fiction and scientific discovery is a story that unfolds over time, often blurring the lines between our wildest dreams and what we can actually make happen. When we look up at the night sky, we aren't just thinking about the distant stars and their planets; we're also reminded of the incredible stories that have sprung from human imagination. These tales, filled with wonder and creativity, have inspired real scientific efforts that have taken us far beyond our home on Earth.

Let's consider the groundbreaking work of Jules Verne. Back in 1865, Verne wrote "From the Earth to the Moon," a story that didn't just captivate readers but also laid the foundation for real space travel. His imaginative depiction of a giant cannon launching a projectile to the moon sparked excitement that went far beyond the pages of his book. Verne took time to carefully explain the mechanics and challenges of traveling to space, painting a lively picture of what a journey to the moon might look like. This wasn't just fantasy; it was a vision that inspired scientists and engineers for generations.

Verne's story was more than just a fun adventure; it was an early example of what we now call engineering feasibility studies. His imagined cannon, the Columbiad, and his detailed calculations about trajectory and gravity inspired early rocketry pioneers. It's mind-blowing to think that nearly 100 years later, in 1969, the Apollo 11 mission made Verne's dreams come true when humans finally walked on the moon. The impact of Verne's imagination echoed through time, turning whimsical fiction into a real achievement. This transformation highlights not only the power of storytelling but also how creative ideas can inspire scientists to turn dreams into plans and plans into groundbreaking accomplishments.

We can't forget to mention Arthur C. Clarke, whose contributions seamlessly blended imagination with scientific exploration. Clarke famously outlined the idea of geostationary satellites in his 1945 essay "Extra-Terrestrial Relays." At the time, it might have seemed like a stretch, yet his vision blossomed into the communications satellites that now orbit our planet. These satellites have revolutionized global communication, weather forecasting, and countless technologies that define our modern lives. Clarke's insight, grounded in scientific understanding and driven by a vivid imagination, beautifully illustrates how the arts and sciences can work together to drive technological progress.

As we explore the world of science fiction, we can't overlook modern works that continue to bridge the gap between fiction and reality. Take Andy Weir's "The Martian," for example. It has captured the hearts of a new generation and sparked conversations about the possibility of humans colonizing Mars. Weir's careful attention to scientific detail—each calculation and every challenge faced by the protagonist, Mark Watney—invites readers to engage deeply with the science of surviving on another planet. The book not only entertains but also inspires real discussions among scientists about the practicality of Mars missions, motivating organizations like NASA to think seriously about sending humans to the Red Planet.

"The Martian" encourages us to ponder what it means to be human in unknown territory. It dives into themes of ingenuity, resilience, and the human spirit's ability to adapt and innovate, even when the odds seem stacked against us. Just like Verne and Clarke, Weir has sparked a renewed interest in exploring other planets, showing us how imagination and scientific investigation can come together to push humanity toward new horizons.

Moreover, the relationship between science fiction and scientific innovation isn't just about individual stories; it thrives in collaborative environments. Today, scientists, engineers, and writers often join forces to explore the unknown. This teamwork is especially critical in fields like astrobiology, where researchers look into the potential for life in extreme environments. The creative storytelling found in science fiction provides valuable frameworks for scientists to consider the different forms life might take on Mars, Europa, or distant exoplanets.

For instance, think about the ongoing search for extremophiles—organisms that thrive in conditions once thought impossible for life. Science fiction creates a space where scientists can share their ideas and explore scenarios that challenge our understanding. These imaginative stories help us ask questions like: How might life adapt in the hidden oceans of icy moons? What could alien organisms look like on the molecular level? These creative explorations feed back into scientific research, encouraging scientists to consider possibilities they might not have thought about otherwise. The connection between storytelling and science highlights the importance of speculative fiction as a powerful tool for broadening our understanding of the universe.

While the stories of the past have certainly shaped scientific exploration, today's science fiction is evolving right alongside technological advancements. The rapid pace of innovation has sparked a resurgence in speculative storytelling, with modern authors diving into themes of artificial intelligence, genetic engineering, and the ethical dilemmas that arise from these technologies. Shows like "Black Mirror" and "Westworld" challenge us to think critically about the impacts of technology on our lives, urging us to confront the consequences of our creations. These narratives don't just entertain; they serve as cautionary tales that prompt us to reflect on the choices we make as we move into an uncertain future.

The bond between science fiction and scientific advancement remains lively, with each one continually inspiring and influencing the other. It's important to see that the ideas presented in fictional works don't exist in isolation; they resonate within the scientific community, pushing researchers to explore new paths and expand the limits of what's possible. The growing public interest in popular science fiction narratives drives funding, research initiatives, and technological development, showcasing how storytelling can significantly impact society's goals.

As humanity stands on the brink of new explorations—be it the search for life beyond Earth or the colonization of other planets—embracing the imaginative stories of today will help guide us through the challenges ahead. The value of creative thinking cannot be overstated. It encourages scientists to dream big, envision solutions that might not yet be obvious, and imagine futures that inspire hope and wonder.

The stories we share about our place in the universe are not just reflections of our fears but also representations of our dreams.

They light the way forward, encouraging us to ask deep questions that get to the heart of existence. The relationship between science fiction and scientific discovery teaches us that creativity can stretch into the realm of what's possible. This connection nurtures a spirit of inquiry that is at the heart of human progress.

As we think about the intertwined paths of fiction and science, it becomes clear that we are participants in this grand cosmic story, not just observers. We shape our future through the tales we tell and the innovations we pursue. The legacy of science fiction resonates in the world of scientific inquiry, inspiring us to push beyond our limits and reach for the stars. In this ever-changing dialogue between imagination and reality, the universe calls to us—inviting exploration, discovery, and dreams that go beyond what we know. Our journey continues, fueled by the limitless potential of human creativity, and as we step confidently into the unknown, we find ourselves at the thrilling crossroads of fiction and scientific discovery.

## The Evolution of Alien Imagery

The way we imagine aliens in science fiction has changed dramatically over the years, showing how our feelings about the unknown have grown and shifted. What once started as scary monsters has transformed into more thoughtful and relatable portrayals that can even make us feel compassion. These changing images of aliens do more than just tell a story; they also reflect our society's values, fears, and hopes.

In the early days of science fiction, aliens were often shown as frightening and suspicious. The creatures that came to life in the pages of magazines or early movies were typically hostile and dangerous. A great example is the classic 1951 film "The Day the Earth Stood Still." In this film, the alien Klaatu arrives on Earth with a strong warning about humanity's destructive ways. However, Klaatu's initial look—a tall, intimidating figure with a menacing robot at his side—creates a sense of fear. While the film carries a hopeful message, it wraps that message in a frightening depiction of the alien. Klaatu represents the worries of the Cold War, a time when the unknown—be it a foreign enemy or the constant threat of nuclear war—instilled fear in people's minds.

During the Cold War, with its prevailing tension and paranoia, we saw a flood of alien invaders in movies that mirrored the fears of that era. Films like "War of the Worlds," based on H.G. Wells' timeless novel, feature Martians landing on Earth with a ruthless desire to conquer. These stories tapped into a deep-rooted fear of the unknown—of forces we couldn't control or fully understand. The aliens were often shown as monstrous beings, lacking any emotion and only wanting destruction. In this way, they were not just invaders; they represented the psychological invasion of fear into everyday life.

As society moved into the 1970s and 1980s, the way aliens were portrayed began to change. The rise of environmentalism and a growing awareness of our impact on the planet opened the door to more complex alien characters. Films like "Close Encounters of the Third Kind" and "E.T. the Extra-Terrestrial" introduced viewers to creatures who were not just threats but also symbols of peace and understanding. In "E.T.," the little alien is gentle and vulnerable, yearning for connection and ultimately becoming a beacon of compassion. This film struck a chord with audiences, reflecting a desire for harmony and empathy in a world that felt increasingly divided. Suddenly, the alien experience was no longer all about fear; it had become a story about empathy and our shared humanity.

This trend toward more heartfelt portrayals continued into the 21st century. Movies like "Arrival," directed by Denis Villeneuve, represent a significant change in how aliens are depicted. In "Arrival," the aliens, known as Heptapods, are first met with fear, but as the story unfolds, we see their true nature. This gradual revelation highlights the importance of communication and understanding when faced with the unknown. The Heptapods, with their unique way of speaking and their different perception of time, challenge our ideas and showcase the limits of human thinking. The film encourages us to ponder deep questions about language, memory, and existence. The Heptapods are not just threats; they are catalysts for personal growth and societal transformation.

The evolution of alien imagery also helps us challenge and rethink existing stereotypes in our own society. As aliens are portrayed in more nuanced ways, it gives us a chance to reflect on how we view those who are different, whether they come from diverse cultures or belong to marginalized groups within our communities. How we portray aliens can shape how people think about diversity, acceptance, and what it means to share a space with those who are fundamentally different from us.

This exploration isn't confined to movies; it also extends deeply into literature. Authors like Octavia Butler and Ursula K. Le Guin dive into the complexities of alien life, tackling themes of colonialism, identity, and coexistence. Butler's "Kindred," although not featuring traditional aliens, uses time travel to explore feelings of alienation and dislocation, pushing readers to face uncomfortable truths about history and power. Likewise, Le Guin's "The Left Hand of Darkness" examines gender and identity through the lens of an alien society, challenging readers to rethink their ideas of what it means to be human.

These literary journeys resonate with current discussions about diversity and inclusion. When science fiction portrays aliens as complex beings with their own cultures, struggles, and hopes, it encourages us to reflect on our own views of acceptance and diversity on Earth. The way we depict the "other" in stories about aliens can serve as a powerful lens on the complexities of human relationships, prompting us to confront our biases and assumptions.

Moreover, the evolution of alien imagery has sparked broader conversations about the future of humanity. As we face urgent global challenges like climate change, social inequality, and rapid technological growth, science fiction provides a rich ground to imagine possible futures. By portraying aliens as guardians of the universe—as seen in films like "Avatar"—these stories stress the importance of caring for our environment and taking collective responsibility. They push us to rethink how we relate to our planet and to one another, fostering a sense of unity as we tackle existential problems together.

As the way we portray aliens continues to evolve, it mirrors the ever-changing landscape of human values and dreams. The stories we tell about extraterrestrial life reflect our deepest hopes and fears, inviting us to engage with the unknown in ways that challenge our understanding of existence. In a time when technology and scientific exploration push the limits of what's possible, imagining aliens can inspire us to dream of futures that go beyond our current boundaries.

Ultimately, the journey of alien imagery in science fiction isn't just about the aliens; it's about us—our values, fears, and dreams. The narratives we create around extraterrestrial beings shine a light on our longing for connection, understanding, and acceptance. They challenge us to face the unknown, encouraging us to find common ground and shared experiences in a universe that can often seem vast and isolating.

Looking ahead, the way we portray aliens will undoubtedly keep evolving, influenced by the shifting tides of human thought and culture. The stories we share about these beings will not only reflect our fears of the unknown but also highlight our ability to empathize and understand one another. In doing so, they hold the potential to inspire hope and possibility, reminding us

that, despite our differences, we are all part of
the same cosmic tale—a tale that welcomes
exploration, curiosity, and ultimately,
connection with the universe and each other.

# Chapter 9: The Consequences of Discovery

The idea of life beyond Earth has fascinated us for centuries. It sparks our imaginations, fuels debates, and challenges our understanding of what it means to exist. When we think about discovering extraterrestrial life, it brings up profound questions about our humanity and our place in the universe. Each new scientific discovery adds to this ongoing conversation, raising the stakes higher than ever. Just the possibility of life beyond our planet could shake up our beliefs, ethics, and the very structures of our society.

In stories and films, encounters with aliens often reflect our hopes and fears. Think about H.G. Wells' "The War of the Worlds," where aliens invade with a menacing presence, or the touching friendship between humans and extraterrestrials in "E.T. the Extra-Terrestrial." These narratives show how we grapple with the unknown. Ultimately, they pose a big question: How would we react if intelligent life were discovered beyond Earth?

Imagine the moment we receive a signal from another civilization. The news would spread like wildfire, lighting up social media and news outlets everywhere. People would feel a mix of excitement and fear. What would our response be? Would we come together as a global community, or would existing divisions grow even wider? This potential contact could spark intense discussions among scientists, theologians, and everyday folks, all trying to understand what it means to have cosmic neighbors.

The philosophical impact of such a discovery could reach far beyond just our initial reactions. For ages, humanity has seen itself as the center of existence, feeling a sense of superiority. Finding other forms of life might force us to rethink this viewpoint. Would we be willing to extend our compassion to non-human intelligences? Would we welcome them with open arms, or would fear and suspicion lead us to see them as threats?

Religious groups would also need to wrestle with what extraterrestrial life means for their beliefs. For many people, religion offers understanding about life, purpose, and the divine. Discovering intelligent life could challenge long-held convictions and spark deep theological debates. Questions about creation, humanity's role in the universe, and the possibility of different paths to salvation would take center stage. Would faith adjust to this new reality, or would it struggle under its weight?

As we think about these philosophical issues, we can't overlook the social implications of such a monumental discovery. Our society is built on norms, values, and shared beliefs that shape how we interact with each other and the world. Discovering extraterrestrial life could shake things up, challenging our ideas about identity and belonging. Would we redefine our communities to include beings from another world, or would we retreat into our own divisions, clinging to old boundaries in the face of the unknown?

Media will play a big role in shaping our reactions to this discovery. The potential for sensationalism is huge, and the way it's reported could either cause panic or promote unity. We'd likely see documentaries and news specials pop up, each offering different takes on what extraterrestrial life means for us. Suddenly, science fiction might feel more like a documentary, as conversations about contact, coexistence, or even conflict spill over into our daily lives—from dinner tables to classrooms.

The scientific community would also undergo a significant shift. Researchers who have devoted their lives to searching for extraterrestrial intelligence would suddenly be in the spotlight. Research priorities would change, funding would be redirected, and fresh collaborations would spring up. The enthusiasm among scientists would be contagious, breathing new life into fields like astrobiology, cosmology, and exoplanet studies. The pursuit of knowledge would accelerate, leading to innovations as we strive to understand the principles that govern life throughout the universe.

But with this scientific excitement comes the need for global cooperation. The search for extraterrestrial life is truly a worldwide effort. A discovery of this scale would cross borders, calling for a unified approach to understanding its implications. How do we handle communication with potential extraterrestrial civilizations thoughtfully and ethically? How do we share this information responsibly? These questions highlight the necessity for collaboration among nations, scientists, and philosophers as we navigate this extraordinary new reality.

In a world already filled with division, the discovery of extraterrestrial life might just spark positive change. It could remind us of our shared humanity, encouraging greater understanding and empathy across cultures. Perhaps it would inspire us to tackle pressing global issues like climate change, poverty, and inequality, recognizing that we all share this cosmic neighborhood. The interconnectedness of life on Earth might serve as a powerful reminder that we are part of something much larger.

Of course, there's a chance that our reactions could reflect humanity's less admirable tendencies. Fear of the unknown could lead to xenophobia—distrust of the "other," whether they come from the stars or just across the border. The challenge lies in fostering a mindset that embraces curiosity and openness instead of retreating into fear and hostility. Education will be crucial in shaping how we respond to new possibilities, equipping people with the tools to think critically and compassionately.

As we consider the societal and philosophical implications of discovering extraterrestrial life, we also need to think about the emotional and psychological effects on individuals and communities. Realizing we're not alone in the universe could be exhilarating but also overwhelming. For some, it might inspire a sense of wonder or a spiritual awakening. For others, it could trigger existential crises, prompting deep reflection on life, purpose, and mortality.

The vastness of the cosmos is both awe-inspiring and intimidating. While some find comfort in their solitude, the idea of sharing this vastness with other sentient beings can challenge our very sense of self. How do we balance our individuality with the knowledge that we are just one species among many? These questions are complex and require careful thought and discussion.

As we explore these intricate reactions, we must also pay attention to the stories we tell about extraterrestrial life. How we frame our cosmic neighbors will shape public understanding. Will they be seen as potential allies, threats, or perhaps as reflections of our own strengths and weaknesses? The narratives we create will shape our collective mindset, guiding how we choose to act in the face of the unknown.

It's vital to remember that our search for meaning and connection goes beyond just looking for extraterrestrial life; it permeates our interactions with one another and the world around us. How we respond to the possibility of life beyond Earth may reveal much about who we are as a species. In a universe full of possibilities, discovering extraterrestrial life could offer a chance for growth, reflection, and a new understanding of our role in the grand scheme of things.

The philosophical discussions that arise from this discovery could lead to a new wave of thought, inspiring humanity to explore fresh ways of understanding existence. As we grapple with being part of a larger cosmic community, we may feel motivated to build a more inclusive, compassionate, and sustainable world right here on Earth. The questions raised by the existence of extraterrestrial life may urge us to value all forms of life, prompting us to protect and cherish our planet and its inhabitants.

Ultimately, how we react to the discovery of extraterrestrial life will be as varied as humanity itself. Each person will interpret this revelation through the lens of their own experiences, beliefs, and values. Some may welcome this adventure with open arms, while others may approach it with skepticism or fear. The beauty of this journey lies in the rich mix of thoughts, emotions, and reflections that will emerge as we face one of the most profound questions in human history: Are we alone in the universe?

As we stand on the brink of this potential discovery, the societal and philosophical reactions will shape not only how we see ourselves but also how we connect with the cosmos and the myriad forms of life that inhabit it. The quest for understanding goes on, and as we move into the unknown, we carry with us the hopes and dreams of generations, eager to connect with the vast universe awaiting us beyond the stars.

## The Role of International Cooperation

The cosmos is a vast and breathtaking realm, filled with galaxies, stars, and the exciting possibility of life beyond our own blue planet. For centuries, humanity has stared into the night sky, pondering the mysteries of the universe and our place within it. As we begin to explore whether we are truly alone in this enormous expanse, the importance of international cooperation shines through. Finding extraterrestrial life would not just be a groundbreaking scientific achievement; it would require a united response from our global community. A disjointed approach could lead to misunderstandings, fears, and conflicts, showing us just how crucial collaboration and shared decision-making are when faced with the unknown.

To understand why international cooperation is so vital for exploring outer space, we need to look back at our history. Humanity's journey into the cosmos has been shaped not only by our thirst for knowledge but also by the conflicts and agreements we've experienced on Earth. After World War II and during the Cold War, there was a growing realization that we needed to work together to tackle global challenges. This urgency led to the creation of the Outer Space Treaty in 1967, a landmark agreement that set the stage for peaceful exploration of space and established key principles for international collaboration. Signed by over a hundred countries, this treaty emphasized that space belongs to all of humanity and that activities in outer space should benefit everyone, regardless of their nationality.

The Outer Space Treaty also reflects a broader commitment to preventing conflict and militarization in space—a concern that is still very relevant today. As nations build their space capabilities, the chances for miscommunication or misunderstandings increase. The idea that contact with extraterrestrial civilizations might be viewed with suspicion or hostility highlights the need for established frameworks like the treaty. If we were to make a discovery, these existing protocols would help guide our collective response, promoting teamwork among nations instead of sowing discord.

In addition to the Outer Space Treaty, several other treaties and international agreements have emerged that illustrate humanity's desire for cooperative space exploration. For instance, the Rescue Agreement of 1968 offers guidelines for helping astronauts in distress, emphasizing that assistance must be provided regardless of nationality. Similarly, the Liability Convention of 1972 outlines principles for addressing damage caused by space objects. Together, these frameworks show a shared commitment to making our explorations of the cosmos safe and accountable.

Even with these important milestones, one key question remains: Are these treaties enough to prepare us for the day we might discover intelligent extraterrestrial life? As we stand on the edge of this possibility, it's crucial to examine existing guidelines like those proposed by the Search for Extraterrestrial Intelligence (SETI). Established in 1984, the SETI protocols provide a structured approach for responding to potential contact with intelligent extraterrestrial life. They cover everything from verifying a signal to considering the broader implications of announcing contact to the public. Most importantly, they call for a global response, emphasizing collaboration among scientists, ethicists, diplomats, and others to engage in thoughtful dialogue and consensus-building.

At the heart of these protocols is a significant ethical question: How do we communicate with a species that is entirely new to us? The guidelines make it clear that scientists must carefully consider how to reach out, ensuring that any communication is responsible and well-thought-out. The potential impact of reaching out to another civilization could be immense. Therefore, a collaborative approach—one shaped by a shared understanding of risks, ethics, and scientific inquiry—is absolutely necessary. Our attempts to connect with another civilization should reflect a commitment to peace, understanding, and respect.

While we do have existing frameworks, the need for global unity is vital in this context. The stakes are high, and the chances for international conflict are substantial. In a world where national interests often take center stage, the discovery of extraterrestrial life could heighten existing tensions or spark new rivalries. Countries may feel compelled to dominate or claim control over extraterrestrial findings, potentially leading to strife instead of collaboration. To prevent this, we must prioritize global unity.

One of the biggest challenges we face is putting aside our divisions to pursue a shared goal: understanding and possibly communicating with extraterrestrial life. The idea that nations, cultures, and communities can come together in the face of the unknown is both daunting and inspiring. Achieving this unity will demand a new mindset—a readiness to set aside differences and embrace our shared humanity as we navigate this uncharted territory.

Imagine a scenario where we receive a signal from another civilization. The immediate reactions would likely vary widely. Some might see this opportunity for dialogue as a chance for unity, while others could view it as a threat. The potential for fear, misunderstanding, or even panic is very real. This is why collaboration becomes even more crucial; countries must come together to develop a unified response that focuses on mutual understanding.

The call for teamwork goes beyond just governments; it also invites scientists, philosophers, and everyday people to get involved. Global discussions will be key to shaping our responses and ensuring a wide range of perspectives are heard. The scientific community, in particular, must lead by example, promoting transparency and sharing information openly. By doing this, we can build trust among nations, which is essential for tackling the complexities of possible contact with extraterrestrial intelligence.

Moreover, education will play a significant role in nurturing global cooperation. As the conversation about extraterrestrial life grows, we must ensure that people everywhere have the knowledge and skills to engage thoughtfully in these discussions. Educational initiatives that foster scientific literacy and critical thinking will be crucial, empowering individuals to navigate this complex topic. By encouraging a culture of curiosity and open-mindedness, we can inspire a spirit of collaboration that helps us face the challenges ahead with resilience and determination.

While the idea of contacting extraterrestrial civilizations can spark excitement and curiosity, it also reminds us that fear of the unknown can drive humanity apart. Our history is filled with examples of xenophobia and misunderstanding stemming from the "other." As we explore this new frontier, we must strive to cultivate an inclusive mindset that embraces curiosity rather than suspicion. To do this, we need to engage in dialogue and promote empathy across cultures and borders.

You might ask: how can we ensure that our response to extraterrestrial life encourages unity instead of division? This question deserves careful thought, as it touches on the very essence of who we are as individuals and communities. The stories we tell about extraterrestrial life will shape public understanding and influence our actions. By portraying extraterrestrial beings as potential allies or partners in our quest for knowledge, we can create a sense of shared purpose that goes beyond national boundaries.

As we navigate these challenges, we must also recognize the role of philosophical discussions in shaping our collective response.

The questions raised by discovering extraterrestrial life go beyond science; they challenge our understanding of existence, identity, and morality. Engaging in these conversations will help us explore our shared humanity and work toward a more compassionate and inclusive society.

The potential discovery of extraterrestrial life could act as a catalyst for global change, urging nations to confront pressing issues such as climate change, poverty, and inequality. Realizing that we are part of a larger cosmic community might inspire us to tackle these challenges with renewed energy. The interconnectedness of life on Earth serves as a powerful reminder that we share a common fate and a responsibility to protect our planet for future generations.

As we think about the implications of potential contact with extraterrestrial civilizations, it's important to consider the emotional and psychological effects such a discovery might have on individuals and communities. For some, the knowledge that we are not alone in the universe could spark feelings of wonder and awe. For others, it might lead to existential concerns or fears about what lies ahead. As a global community, we need to be ready to support one another through these complex emotions, helping individuals process these profound realizations.

In the face of uncertainty, storytelling will also play a crucial role. How we tell our stories about extraterrestrial life will shape public perception and understanding. The narratives we create—whether through books, films, or media—will influence how we think about our cosmic neighbors. As we shape these stories, we must ensure they inspire curiosity and openness instead of fear and division.

Ultimately, the role of international cooperation in responding to the discovery of extraterrestrial life isn't just a matter of governance; it reflects who we are as a species. As we stand on the brink of this significant discovery, we must realize that our responses will be shaped by our values, beliefs, and collective dreams. The journey toward understanding the cosmos requires humility, empathy, and a steadfast commitment to working together.

In a universe filled with possibilities, discovering extraterrestrial life could be a turning point that brings humanity together in the pursuit of knowledge and understanding. By fostering collaboration and shared purpose, we can navigate the complexities of this incredible new reality, embracing opportunities for growth, reflection, and connection. The very act of seeking to understand other forms of life may lead us to a deeper understanding of ourselves—a powerful reminder that we are all part of a vast and interconnected web of existence.

As we step into this unknown territory, let us carry with us the hopes and dreams of generations who have gazed at the stars, yearning for connection. By encouraging international cooperation, we can ensure that our journey into the cosmos not only leads to discoveries beyond our wildest dreams but also to a more united, compassionate world right here on Earth. The cosmos awaits, and it's up to us to approach it with open hearts and minds, ready to embrace the wonders and challenges that lie ahead.

## What Insights Might Alien Life Offer?

The idea of alien life has always sparked our curiosity and imagination, blending science with our wildest dreams. What if we are not alone in the universe? What if, hidden among the countless stars and galaxies, there are life forms that could challenge and deepen our understanding of existence? Discovering alien life could change everything, affecting not just science but also our philosophical beliefs and the future of humanity.

Are We Alone In The Universe?

When we think about what we might learn from studying alien life, a world of endless possibilities opens up. Imagine finding life forms thriving in places we once thought were impossible for life to exist. Extremophiles—those remarkable organisms that can survive extreme conditions right here on Earth—give us a glimpse into the wonders that could be out there. From the scorching hot geysers of Yellowstone National Park to the icy depths of Antarctica, extremophiles have shown us how life can adapt to extreme heat, intense pressure, and complete darkness. Their resilience highlights the creativity of life, a trait that might also be found in extraterrestrial beings.

Take the tardigrade, often lovingly called the "water bear." This tiny creature can withstand harsh radiation, extreme dryness, and temperatures ranging from almost absolute zero to over 300 degrees Fahrenheit. Studying these organisms could profoundly change how we view biology, evolution, and the possibility of life beyond our planet. The existence of extremophiles pushes us to rethink our definitions of life itself. If life can thrive in conditions we once thought were deadly, what might we find on planets with sulfuric acid in the atmosphere or in the icy oceans beneath moons like Europa? The answers could not only reshape scientific understanding but also lead to exciting technological advancements.

Now, think about the incredible potential for advancements in biotechnology that could come from alien life. If we discover extraterrestrial organisms with unique biochemical processes, they might unlock new breakthroughs in medicine, energy, and agriculture. Scientists on Earth have already turned to nature for inspiration—this practice is called biomimicry, where we mimic the strategies evolved by organisms to solve human problems. Finding alien biology could take this to a whole new level, giving us new ideas for sustainable practices and fresh solutions to long-standing issues. Who knows what secrets could be hidden in the DNA of alien life that might transform our medical treatments or lead to more efficient energy systems?

But the scientific excitement surrounding the possibility of alien life is only part of the story. The philosophical implications of such a discovery prompt us to deeply consider consciousness, intelligence, and what it means to exist. For centuries, people have pondered profound questions: What does it really mean to be alive? Are we mere accidents in a vast universe, or is our existence part of something greater? Discovering alien life could reflect our beliefs back at us while also amplifying the questions that define our humanity.

Imagine if we were to find intelligent alien civilizations; it would challenge everything we think about consciousness. For a long time, we've believed we're at the top of the cognitive ladder, assuming our intelligence is unique. What if we encounter beings with forms of intelligence completely different from ours? This realization could broaden our understanding of consciousness, encouraging us to see that intelligence may take many forms, shaped by the specific environments and challenges faced by different life forms. The conversations sparked by such encounters could enrich our understanding of our own minds, creativity, and ability to empathize.

Moreover, meeting alien intelligence would push us to think about our values and ethics. How do we treat beings from another world? Would we show them the same respect and dignity we demand for ourselves? Reflecting on our history of conflict and colonization on Earth, we need to be cautious about ethical missteps in our interactions with alien life. The challenge lies in creating a framework for engagement that truly embodies compassion, equality, and understanding. This new perspective on our place in the universe could lead to deeper discussions about our moral responsibilities to one another and to any life forms we might encounter, regardless of where they come from.

As we start to entertain the possibility of contacting extraterrestrial civilizations, we should also imagine the future possibilities such encounters could bring. Just picture a collaborative effort across the cosmos, sharing knowledge and resources to tackle challenges that affect all of us. The idea of humanity joining forces with intelligent alien beings could inspire a sense of unity and purpose that we've often struggled to find here on Earth. This cosmic community could create a narrative that prioritizes cooperation over competition, encouraging us to build connections that go beyond our planet's boundaries.

While collaboration may sound idealistic, it's rooted in the reality that we're facing common challenges. The environmental crises affecting our planet, the struggle for fair access to resources, and the quest for a sustainable future are issues we can't solve alone. In this imagined future, contact with alien civilizations could bring insights and technologies that help us tackle problems that have stumped humanity for ages. The chance for mutual enrichment is exciting, offering not just solutions but opportunities for cultural exchange that could broaden our perspectives in ways we can't yet imagine.

However, as we explore these possibilities, we must be mindful of the complexities and potential challenges of such interactions. How do we establish communication and collaboration that respect our diverse backgrounds and beliefs? Navigating the nuances of interstellar diplomacy would require a level of cultural understanding and respect that we're still striving to achieve on our own planet. Perhaps the key is to foster a global mindset—nurturing empathy, open-mindedness, and flexibility to connect with the unknown.

In thinking about the future, we also need to address the philosophical implications of being part of a larger cosmic community. Understanding that we belong to a bigger picture could foster a sense of connection that goes beyond individual or national identities. This realization might reshape how we see ourselves, urging us to look beyond our immediate surroundings and embrace the idea that we are all part of a universal family. Such a shift could inspire a sense of responsibility for the well-being of not just our planet, but for the cosmos as a whole—prompting us to become stewards of life, whether it's here on Earth or beyond.

As we stand on the brink of possibility, the thought of discovering extraterrestrial life is both thrilling and intimidating. It encourages us to expand our perspectives, to question our assumptions about intelligence and existence, and to welcome the unknown. The scientific, philosophical, and future-focused insights we gain from this journey could transform how we understand ourselves and our place in the cosmos.

In this exploration, storytelling becomes crucial. The tales we tell about alien life will shape our expectations and actions. As we imagine the possibilities, let's weave our stories with hope, curiosity, and the potential for unity. The narratives we create could inspire future generations, encouraging them to look up at the stars and wonder about the lives that thrive beyond our own world.

As we consider the profound questions that might arise from discovering alien life, it's important to be ready to navigate the emotional landscape that such revelations could bring. Feelings like wonder, fear, hope, or even despair may surface, and it's vital to acknowledge and understand them. Supporting each other as we process these emotions will be key as we face a reality that might fundamentally change our understanding of existence.

Ultimately, the journey to uncover what insights alien life might offer is an invitation to open our minds and hearts. It challenges us to embrace the uncertainty, engage in meaningful conversations, and nurture a spirit of collaboration that crosses the boundaries of our world. The universe is vast and filled with potential discoveries, and as we gaze toward the stars, let's hold onto the hope that our quest for understanding can lead us to a kinder and more connected future.

As we reflect on the awe-inspiring possibility of alien life, let's remember that our exploration is not just about seeking the unknown, but also about deepening our understanding of what it means to be human. The answers we seek aren't just in the stars; they lie within us too. It's through the questions we ask and the stories we share that we can uncover the wisdom that guides us through the uncharted territories of existence. As we courageously step into this cosmic frontier, may we do so with open hearts and minds, ready to embrace the insights that await us among the stars.

Miles Kepler

# Chapter 10: The Future Awaits: Continuing the Search

Imagine standing under a vast, star-lit sky on a clear night, with the Milky Way glowing like a shimmering river of light above you. Your mind races as you think about the mysteries of the universe and whether there could be life beyond our fragile blue planet. For centuries, people have wondered if we're alone in this vast cosmos or if, somewhere among all those twinkling stars, other intelligent beings are looking back at us, filled with their own curiosity. Right now, we find ourselves at a special moment in history, where hope and innovation meet, and technology is driving us toward answers we once thought were impossible to find.

The search for extraterrestrial life isn't just a dream from science fiction anymore; it's a serious scientific endeavor, powered by amazing advancements in technology. These innovations have opened up new paths for exploration, letting us dive deeper into the cosmos than ever before. From ground-based observatories to daring space missions, each new technology brings us closer to solving the mystery of life beyond Earth.

One of the most exciting breakthroughs has come in the world of telescopes. The James Webb Space Telescope (JWST), which launched in December 2021, marks a major leap in our ability to observe the universe. With its incredible capacity to capture light from the earliest galaxies and distant exoplanets, the JWST is already changing how we understand the cosmos. Its powerful infrared sensors can examine the atmospheres of exoplanets, looking for chemical signs of life, known as "biosignatures," that might indicate the presence of living organisms. This telescope has not only stretched the limits of our vision but has also sparked a new wave of excitement and hope among scientists.

As we venture further into space, the ways we explore nearby worlds have also grown. Missions to Mars are leading the charge in our exploration efforts. Rovers like Perseverance and Curiosity have been sending back a treasure trove of information. These robotic explorers roam the Martian landscape, gathering samples and running experiments to learn about the planet's geology and climate. The hunt for ancient microbial life on Mars isn't just about discovery; it's also about exploring our own beginnings and learning if life ever existed on our neighboring planet.

Beyond Mars, the icy moons of Jupiter and Saturn, like Europa and Enceladus, are becoming exciting candidates in the search for life. Scientists are especially fascinated by Europa, which is thought to have a hidden ocean beneath its icy surface. The upcoming Europa Clipper mission aims to study this moon in incredible detail, looking into its potential to support life. Could this frozen world, with its vast ocean and organic materials, be home to forms of life that evolved without any connection to the rest of our solar system?

Additionally, developments in astrobiology are giving us fresh insights into the possibility of life in extreme conditions. By examining extremophiles—organisms that thrive in environments once deemed unlivable—researchers are reconsidering what it takes for life to exist. These findings suggest that life could be lurking in places we never thought possible, like the thick atmospheres of gas giants or within the hidden oceans of icy moons.

As we keep gazing at the stars, the technology that supports our exploration goes beyond just advanced equipment and tools. Artificial intelligence (AI) and machine learning are transforming how we handle and analyze data from space missions. The massive amount of information gathered by telescopes and rovers can be overwhelming, but AI algorithms can help sift through this data, spotting patterns and anomalies that might escape human attention. This partnership between human creativity and machine learning is pushing our search for extraterrestrial life into exciting new territories.

While the strides in technology are thrilling, they also raise important questions about the ethics of our exploration. What responsibilities come with our desire to explore and possibly settle on other worlds? As we look to Mars and beyond, we need to think carefully about how our presence might affect these celestial bodies. The risk of contaminating pristine environments raises valid concerns about keeping extraterrestrial ecosystems intact. Just as we learn to protect our own planet, we should extend the same care to other worlds—whether they are barren rocks or places that could support life.

Moreover, the chance of finding life beyond Earth brings with it profound moral questions. How we interact with other intelligences—whether they are tiny microbes or advanced civilizations—will require us to rethink our ethical principles. What rights might non-human beings have? What obligations do we have to protect their habitats and ensure we treat them with respect? These aren't just theoretical discussions; they are crucial conversations we need to have as we prepare for the next steps in our cosmic journey.

The search for extraterrestrial life reflects some of humanity's deepest instincts: the urge to explore, to understand, and to connect. Whether we eventually find that we are alone in the universe or that we share it with other sentient beings, the journey itself is transformative. It pushes us to think about our place in the grand scheme of things, fostering a sense of humility as we realize how small our worries are in the vastness of space.

As we look ahead, collaboration will be key. The quest for life beyond Earth isn't just for scientists and researchers; it calls for the combined imagination of all humanity. Artists, writers, and creative thinkers from every corner of life can add their voices to this story, helping to build a culture of curiosity and wonder that inspires future generations.

In the coming years, as technology continues to evolve at a rapid pace, we can expect new missions and discoveries that stretch the limits of our knowledge. Each new finding brings us closer to answering that age-old question: Are we alone in the universe? And as we reach for the stars, we also need to look inward, reflecting on our values and responsibilities not just for our own planet but for the cosmos beyond.

This journey is just beginning, and the innovations in technology and exploration are the fuel that will drive us forward. As we focus on the future, we become part of a larger story—one that goes beyond individual lives and unites us in the shared quest for knowledge, understanding, and the opportunity for connection. So, let's embrace this adventure with open hearts and curious minds, ready to explore the endless expanse that lies ahead. The future is indeed bright.

## Humanity's Aspirations for Colonization

Imagine a future where humanity has broken free from the limits of Earth, reaching out to the stars with a shared sense of purpose that brings us together. A world where the idea of colonizing other planets isn't just a fantasy, but a real goal pursued by passionate scientists, engineers, and dreamers. Mars, often called the Red Planet, is our main target in this cosmic adventure—a barren land that has sparked our imaginations for generations. As we look toward this celestial neighbor, we need to think about the ambitions, advantages, and ethical responsibilities that come with our desire to colonize.

There's a buzz in the air about the idea of humans settling on Mars. NASA,

SpaceX, and various international partnerships are all working hard to establish a human presence on the Martian surface. With dreams of manned missions in the next decade, these organizations are making what once felt impossible a reality. NASA's Artemis program, which aims to bring humans back to the Moon, is seen as a stepping stone on the way to Mars. By improving our technologies and understanding long-term space travel, we can gather important knowledge that will support future missions on the Red Planet. At the same time, SpaceX, led by the ambitious Elon Musk, hopes to send the first crewed missions to Mars as early as 2026 with its Starship spacecraft. Musk's vision includes creating a self-sustaining city on Mars, turning it into a "backup" for humanity.

    The reasons for wanting to colonize Mars go beyond just exploration; they reflect our deep desire to survive, grow, and push our boundaries. In a world filled with serious threats like climate change, pandemics, and potential asteroid impacts, the idea of Mars as a second home becomes more urgent. As we face the fragile nature of our existence, having a foothold on another planet gives us hope that humanity can persevere. The dangers of staying on just one planet are clear, making the goal of a Martian colony feel like not just an adventure, but a necessary step to protect ourselves against the unpredictable nature of life on Earth.

Colonizing Mars could also lead to amazing scientific benefits. Establishing a presence on the Red Planet would open up new opportunities for research and technological advancements. Think about the discoveries we could make regarding planetary formation, geological processes, and even the possibility of life beyond Earth. Mars could serve as a unique lab to study climate change on a much larger scale, offering insights that could help us understand our own climate crisis. Additionally, the technologies developed for sustaining life on Mars—like advanced life support systems, renewable energy solutions, and efficient resource management—could greatly benefit sustainable living back on Earth. The innovations we create in the tough Martian landscape might improve the lives of people all over our planet.

But the dream of colonization isn't without its ethical challenges and potential downsides. As we think about our goals, we need to ponder the moral implications of changing another world. Mars may seem empty, a lifeless sphere just waiting for us to arrive, but what if there are tiny ecosystems yet to be uncovered? Protecting these possible Martian organisms is an important ethical question. If life exists, even in its simplest forms, what rights do those organisms have? Should we approach this new world carefully, respecting its existence, or should we assert ourselves as its new rulers? These questions challenge our moral beliefs and the responsibilities we carry as caretakers of our own planet and beyond.

Creating a sustainable life on Mars is a daunting task. The Martian atmosphere, mostly made up of carbon dioxide, is not suitable for human life. Developing life support systems to provide breathable air, safe drinking water, and food will require groundbreaking innovations. Techniques like hydroponics and aeroponics could be essential for growing crops in Martian soil, and advanced recycling technologies will be needed to reduce waste and make the most of our resources. Moreover, we can't ignore the mental challenges that come with long-term isolation on Mars. As humans make their way into the stark Martian landscape, we need to be ready for the emotional and psychological impacts of leaving the comforts of Earth behind. Issues like cabin fever, anxiety, and depression must be addressed with strong support systems that prioritize the mental well-being of Martian settlers.

Terraforming Mars is another intriguing idea that aims to change the planet into a more Earth-like environment. While the thought of modifying Mars to make it more livable is exciting, it raises serious ethical concerns. Do we really have the right to alter another planet's ecosystem, even if our aim is to make it habitable for ourselves? The idea of playing god with another world is filled with uncertainty. The long-term effects of terraforming could cause unexpected ecological issues, possibly wiping out Martian life forms we haven't even discovered yet. We must let ethical considerations guide our decisions as we navigate the tricky terrain of planetary colonization.

  Additionally, the vast distance and complexities of traveling to Mars pose significant challenges. Getting to Mars, which takes about six to nine months with our current technology, requires careful planning and execution. Spacecraft need to be designed to handle cosmic radiation, the effects of microgravity, and the psychological pressures of long journeys. Keeping astronauts healthy and safe over long periods is a monumental job. Plus, figuring out how to transport supplies and equipment to Mars adds even more layers of complexity to this endeavor. With the technology we have today, maintaining human life on another planet is a complicated and multifaceted challenge.

As we think about our role as potential colonizers of Mars, we also need to reflect on what it means to be responsible caretakers of another world. Our duties go beyond just protecting Martian ecosystems; they involve a deeper commitment to ensuring that our actions don't repeat the mistakes of the past. For instance, the colonization of the Americas led to the destruction of indigenous cultures and ecosystems. As we look to Mars, we have a chance to learn from history and take a more ethical approach to exploration and settlement. The conversation about colonization must focus on respect, humility, and a sincere desire to coexist with any life forms we may find.

Ultimately, the idea of colonizing Mars brings up big questions about our existence and purpose as a species. What drives our never-ending curiosity to explore the unknown? Is it a struggle for survival, a quest for knowledge, or perhaps an inherent need to discover? This urge to push boundaries and seek the extraordinary is what makes us human. As we look up at the stars, we are reminded that we are not just explorers of distant planets, but also guardians of our own Earth.

Standing at the dawn of a new era, our hopes for colonizing Mars offer a chance to bring humanity together in a shared effort.

Conversations about this venture must include diverse voices, promoting dialogue across cultures, fields, and viewpoints. Artists, writers, philosophers, and scientists all have important roles in shaping the story of our future in space. We need to create a vibrant and inclusive vision of what life on Mars could be—a vision that embraces exploration while respecting the integrity of the worlds we might encounter.

The future of human colonization on Mars is not just a matter of technological prowess; it reflects our values, ethics, and ambitions as a species. As we gaze at the stars, we must keep in mind that the choices we make today will shape the future for generations to come. The journey to Mars isn't merely about extending humanity's reach; it's about redefining what it means to be human in a universe filled with possibilities.

In this grand adventure, as we consider the challenges and opportunities of colonization, we find ourselves at a crossroads. The dreams of creating a Martian colony come with questions that dig deep into our shared understanding. As we prepare for this monumental leap into the cosmos, let's foster a mindset that embraces curiosity, accountability, and a united commitment to caring for life, both on Earth and on the distant shores of Mars. The future is here, and it calls us to rise to the occasion, to be the explorers, caretakers, and protectors of both our planet and the universe that lies beyond.

## Ethics of Exploration and Contact

The thought of meeting intelligent life from another world is both exciting and intimidating. It stirs up a mix of feelings—curiosity, a bit of fear, and a strong sense of responsibility. As we search for extraterrestrial life, we're faced with serious ethical questions. What will happen if we reach out to the universe? And what if it responds? This isn't just a fun idea to dream about; it's a question that thinkers, scientists, and philosophers have debated for ages.

To really grasp the ethical issues tied to our quest for extraterrestrial contact, we can look at the history of interactions between different human cultures. Many times, these encounters have shown a complex dance of curiosity and conquest, where genuine interest often led to destruction and exploitation. Take, for example, the European explorers who set out to discover new lands. They were driven by a thirst for knowledge, but too often they left behind a legacy of devastation and the loss of cultures. These lessons are crucial as we think about how we might interact with intelligent alien species. Will we offer a hand of friendship, or will we repeat the mistakes of our ancestors and impose our ways on others?

The ethical issues we face in our search for life beyond Earth are varied and intricate. On one hand, there's the thrilling possibility of making contact, of sharing knowledge and learning from beings that might see the universe in a completely different way. This chance for mutual growth—a partnership between civilizations, if you will—fuels our desire to reach into the cosmos. Yet, we cannot ignore the very real dangers that such contact could bring. A poignant example comes from history: when European explorers unintentionally brought diseases to indigenous populations. It raises a big question: what if we accidentally introduced a harmful alien virus to Earth, or what if we brought microbes that could wreak havoc on another world? The risks are incredibly high, and we need to think carefully about the impact of our actions.

Cultural imperialism is another dark shadow the history of exploration casts over our thoughts about meeting alien civilizations. Human expansion has too often led to the oppression of local cultures and the erasure of their identities. As we consider the idea of reaching out to intelligent life, we must ask ourselves: how can we ensure that we don't force our values, beliefs, or ways of life onto others? The legacy of colonization serves as a warning that the cultures of those we meet must be honored and respected. Our goal should be to build understanding and strengthen coexistence, not to dominate.

The "Prime Directive," a concept made famous by the science fiction series Star Trek, gives us a fascinating way to think about the ethical issues surrounding contact with alien civilizations. This fictional guideline encourages non-interference in the natural growth of alien species. Even though it's from a TV show, the idea carries significant real-world implications. Should we approach contact with a similar commitment to non-interference, allowing other civilizations to grow and flourish on their own? This perspective challenges us to reflect on the responsibilities that come with our advanced technology and knowledge. As we learn more about the universe, we also need to confront the moral obligations we have to protect and nurture any life forms we may find.

The debate about whether we should actively seek contact or take a more cautious stance has captured the attention of ethicists, scientists, and philosophers. Some people believe that reaching out to the stars is a natural extension of our human curiosity, a deep-seated desire to connect with others. Others warn that such efforts might lead to unexpected consequences and suggest a more careful approach instead. Who can truly say what awaits us out there? As we think about our role in the universe, we have to weigh whether the risks of making contact outweigh the possible benefits. Is it our responsibility to explore and connect, or should we stay quiet and observant?

In this conversation, it's important to highlight the voices of those who have dedicated their lives to studying ethics in exploration. Well-respected philosophers and ethicists have shared their thoughts on the responsibilities that come with searching for extraterrestrial life. They remind us that our actions should be guided by principles of respect, caution, and a commitment to the greater good. Should we approach this quest with the eagerness of explorers claiming territory, or should we do so with humility, fully aware of our position in the vastness of existence? These questions are complex and demand thoughtful consideration.

Additionally, we must think about how our search for extraterrestrial life impacts our own planet. In a world often divided by nationality, culture, and beliefs, the quest for life beyond Earth could unite us under a shared mission. What if, through our search for contact with alien civilizations, we discovered the common values that link us as a species? The idea of standing together with fellow humans—scientists, dreamers, skeptics—crossing borders to embrace the potential for collaboration might be one of the most significant outcomes of our cosmic journey.

As we navigate the maze of ethical considerations surrounding the exploration of life beyond Earth, we must remember the inherent value of life, no matter where it comes from. Life, whether it exists on our planet or beyond, is rich and complex and deserves our respect. Our fascination with the cosmos should be guided not just by a desire for knowledge but also by a heartfelt appreciation for the interconnectedness of all living beings. Every organism, no matter how strange it may seem, plays a role in the grand story of existence.

As we think about the ethical challenges of interstellar contact, we find ourselves at a crucial crossroads. This moment calls for reflection, pushing us to examine our motivations, values, and duties. Perhaps the clearest lesson that emerges from these discussions is the need for humility. As we reach out to the stars, we should do so with reverence for the vast universe and the many forms of life it may hold. The cosmos is a complex web of connections, and our place in it should be one of stewardship, not dominance.

Ultimately, the search for extraterrestrial life is more than just a scientific mission; it's a reflection of who we are as a species. Our dreams for discovery and connection are intertwined with our moral obligations to protect and cherish the richness of life, both known and unknown. The questions we ask and the choices we make in our quest to understand the universe will echo far beyond our own planet, shaping the legacy we leave for the generations to come.

As we stand at the edge of this exciting new frontier, the ethical aspects of exploration and contact remind us that our journey into the cosmos is about more than just technology or discovery. It's about the stories we choose to tell, the values we hold dear, and the paths we carve out in the name of understanding. The universe is waiting for us, offering a chance to redefine what it means to be human—a journey that is as much about nurturing our own humanity as it is about reaching for the stars.

*Conclusion*

As we conclude our cosmic journey, we find ourselves back where we started – gazing up at the night sky, filled with wonder and curiosity. Our exploration of the search for extraterrestrial life has taken us through scientific breakthroughs, philosophical ponderings, and cultural reflections.

While we may not have definitive answers yet, our quest continues. Each new discovery, each technological advancement brings us closer to understanding our place in the universe. The search for life beyond Earth is more than scientific pursuit; it's a reflection of our deepest desires to connect and understand our existence.

As you close this book, we hope you carry with you a renewed sense of wonder about the cosmos. Keep looking up, keep questioning, and keep exploring. The universe is vast, and its mysteries are waiting to be unraveled. Who knows? The next great discovery could be just around the corner, forever changing our understanding of life and our place in the cosmic story.

Miles Kepler

www.ingramcontent.com/pod-product-compliance
Lightning Source LLC
Chambersburg PA
CBHW031926240526
45464CB00023B/1698